Negative Capacitance in Ferroelectric Materials

Michael Hoffmann

Technische Universität Dresden

Negative Capacitance in Ferroelectric Materials

M.Sc.
Michael Hoffmann

der Fakultät Elektrotechnik und Informationstechnik
der Technischen Universität Dresden

zur Erlangung des akademischen Grades

**Doktoringenieur
(Dr.-Ing.)**

genehmigte Dissertation

Vorsitzender:	Prof. Dr.-Ing. habil. Christian Georg Mayr	Tag der Einreichung:	30.04.2020
Gutachter:	Prof. Dr.-Ing. Thomas Mikolajick	Tag der Verteidigung:	28.07.2020
	Prof. Dr. rer. nat. Franz Kreupl		

Bibliografische Information der Deutschen Nationalbibliothek:
Die Deutsche Nationalbibliothek verzeichnet diese Publikation in der Deutschen
Nationalbibliografie; detaillierte bibliografische Daten sind im Internet über dnb.dnb.de abrufbar.

© 2020 Michael Hoffmann
Herstellung und Verlag:
BoD – Books on Demand, Norderstedt

ISBN: 978-3-7519-9936-6

Abstract

Negative capacitance in ferroelectric materials is promising for overcoming the fundamental limits of energy efficiency in electronics. In particular, ferroelectrics based on HfO_2 and ZrO_2 are of significant practical interest due to their compatibility with state-of-the-art semiconductor manufacturing processes and excellent thickness scalability. However, so far, progress on negative capacitance devices has been limited due to the use of simplified theoretical models and a lack of experimental data on HfO_2 and ZrO_2 based ferroelectrics. In this work, theoretical and practical aspects of ferroelectric negative capacitance were investigated with a focus on HfO_2 and ZrO_2 based ferroelectrics and the role of domain formation. Ferroelectric and ferroelectric/dielectric thin film capacitors were fabricated and characterized to gain insights into the physics of ferroelectric negative capacitance. Generally, it was found that ferroelectric negative capacitance effects can either be stabilized or transient and intrinsic or extrinsic in nature. Transient negative capacitance was experimentally demonstrated in polycrystalline HfO_2 based and epitaxial perovskite based metal-ferroelectric-metal capacitors with a series resistor. In the latter, transient negative capacitance was shown to be an extrinsic effect caused by domain nucleation and growth dynamics. By connecting a ferroelectric and a dielectric capacitor in series, differential voltage amplification due to transient negative capacitance was demonstrated. However, large hysteresis was found to be inevitable in metal-ferroelectric-metal capacitors which exhibit such transient negative capacitance. Theoretical investigations based on Landau theory suggest that negative capacitance cannot be stabilized in metal-ferroelectric-metal capacitors due to the formation of domains and leakage currents. Therefore, devices with a ferroelectric/dielectric stack are needed to stabilize negative capacitance. Using ferroelectric $Hf_{0.5}Zr_{0.5}O_2$ and dielectric Al_2O_3 or Ta_2O_5 layers, stabilized negative capacitance was observed by using fast pulsed voltage measurements to prevent charge injection and hysteretic switching. 'S'-shaped polarization-electric field characteristics as well as a double-well free energy landscape for ferroelectric $Hf_{0.5}Zr_{0.5}O_2$ was extracted from experimental data in agreement with Landau theory. Nevertheless, it is still unclear whether these negative capacitance effects in HfO_2 and ZrO_2 based ferroelectrics are of intrinsic or extrinsic nature. Lastly, it was proposed and demonstrated that stabilized negative capacitance effects in ferroelectric/dielectric capacitors under pulsed voltage operation are promising for electrostatic supercapacitors with higher energy storage density and efficiency than previously possible.

Kurzzusammenfassung

Negative Kapazität in ferroelektrischen Materialien ist vielversprechend für die Überwindung der fundamentalen Grenzen der Energieeffizienz in elektronischen Schaltungen. Insbesondere HfO_2 und ZrO_2 basierte Ferroelektrika sind von großem praktischen Interesse aufgrund ihrer Kompatibilität zu aktuellen Halbleiterherstellungsprozessen und der Skalierbarkeit ihrer Schichtdicke. Trotzdem wurde der Fortschritt bei Bauelementen mit negativer Kapazität bisher gebremst durch die Verwendung vereinfachter theoretischer Modelle und den Mangel an experimentellen Daten von HfO_2 und ZrO_2 basierten Ferroelektrika. In dieser Arbeit wurden theoretische und praktische Aspekte ferroelektrischer negativer Kapazität mit einem Fokus auf HfO_2 und ZrO_2 basierte Ferroelektrika, sowie die Rolle der Domänenbildung untersucht. Um neue Einblicke in die Physik der ferroelektrischen negativen Kapazität zu bekommen, wurden Kondensatoren mit ferroelektrischen, sowie ferroelektrisch/dielektrischen Dünnschichten hergestellt und charakterisiert. Generell zeigte sich, dass ferroelektrische negative Kapazität entweder stabilisiert oder transient ist und als extrinsischen oder intrinsischen Ursprungs eingeordnet werden kann. Transiente negative Kapazität konnte experimentell in Metall-Ferroelektrikum-Metall Kondensatoren basierend auf polykristallinem HfO_2 und epitaktischen Perowskit-Ferroelektrika durch Verwendung eines Serienwiderstandes nachgewiesen werden. In letzteren konnte aufgrund der Domänennukleations- und -wachstumskinetik gezeigt werden, dass der transiente negative Kapazitätseffekt extrinsischen Ursprungs ist. Durch die Reihenschaltung eines ferroelektrischen und eines dielektrischen Kondensators konnte eine differenzielle Spannungsverstärkung durch transiente negative Kapazität beobachtet werden. Allerdings ließen sich Hysterese-Effekte in solchen Metall-Ferroelektrikum-Metall Kondensatoren mit transienter negativer Kapazität nicht verhindern. Theoretische Analysen basierend auf der Landau-Theorie legen nahe, dass negative Kapazität aufgrund von Domänenbildung und Leckströmen nicht in Metall-Ferroelektrikum-Metall Kondensatoren stabilisiert werden kann. Daher sind Bauelemente mit ferroelektrisch/dielektrischen Schichtstapeln notwendig um die negative Kapazität zu stabilisieren. Unter Verwendung ferroelektrischer $Hf_{0.5}Zr_{0.5}O_2$ und dielektrischer Al_2O_3 oder Ta_2O_5 Schichten, sowie kurzer Spannungspulse, konnte eine stabilisierte negative Kapazität beobachtet werden, wobei Ladungsinjektion und Hysterese-Effekte verhindert werden konnten. 'S'-förmige Polarisations-elektrisches Feld-Kurven sowie eine Doppel-Minimum-Energielandschaft für ferroelektisches $Hf_{0.5}Zr_{0.5}O_2$ konnte in Übereinstimmung mit der Landau-Theorie extrahiert werden. Trotzdem kann noch nicht eindeutig bestimmt werden, ob die gemessene negative Kapazität in HfO_2 und ZrO_2 basierten Ferroelektrika intrinsischen oder extrinsischen Ursprungs ist. Zuletzt wurde gezeigt, dass stabilisierte negative Kapazität in ferroelektrisch/dielektrischen Kondensatoren vielversprechend für elektrostatische Superkondensatoren ist, welche eine höhere Energiedichte und Effizienz aufweisen, als dies bisher möglich war.

Contents

1 Introduction

The capacitor, besides the resistor and the inductor, is one of the fundamental building blocks of any electric or electronic system. From the transmission and storage of energy, over transportation to the processing and storage of information, capacitors have become indispensable in our everyday lives. Indeed, our whole information technology is based on the extremely fast charging and discharging of unfathomable amounts of tiny capacitors e.g. in dynamic random access memories (DRAM) and the gates of metal-oxide-semiconductor field-effect transistors (MOSFETs) representing '0' or '1' as binary logical states. Since capacitors are passive devices, it has long been thought that capacitance must always be a positive quantity. However, as already widely accepted in the context of resistance, passive devices can indeed exhibit negative (differential) resistance under certain conditions, which is used in many practical applications. In contrast, the idea of a (differential) negative capacitance (NC) [1] is relatively new and unexplored, but is promising for overcoming some of the fundamental limitations of energy efficiency in electronics.

In integrated nanoelectronic circuits, the power dissipation is mainly limited by how much voltage is needed to switch a MOSFET between its on- and off-state ('0' or '1'), which is around 0.7 V in current state-of-the-art complementary metal-oxide-semiconductor (CMOS) technologies. Lowering the supply voltage further could drastically reduce the overall power dissipation of such integrated circuits. However, there is a fundamental limit to how much the supply voltage can be reduced in conventional MOSFETs, given by the Boltzmann distribution of electron energies in the source terminal [2]. This means that the supply voltage of future integrated circuits cannot be reduced much further than $\sim$0.5 V, which might be reached relatively soon. Since further increases in device density necessitate a reduction of the supply voltage, this limit might ultimately lead to the discontinuation of Moore's law [3], which states that the density of devices on a semiconductor chip doubles roughly every two years. To continue the dramatic improvements in energy efficiency of electronics enabled by Moore's law in the past decades, new solutions are necessary, one of the most promising of which is NC.

Recently, it was suggested that ferroelectric materials could exhibit NC, which could in principle solve the supply voltage scaling issue described before [4]. A ferroelectric gate insulator in a transistor could act as a step-up voltage transformer, internally amplifying the applied voltage. This original idea was met with considerable skepticism in the beginning, as it was based on a simplified single-domain ferroelectric model and experimental evidence for ferroelectric NC was scarce. Furthermore, since most known ferroelectric materials at the time were not compatible with standard semiconductor manufacturing techniques and had limited scalability in terms of thickness, initial expectations for near-term applications were low. Some researchers have even suggested that the idea of ferroelectric NC is incompatible with the basic physics of ferroelectrics. As it has since turned out, the appearance of NC seems to be intimately related to the origin of ferroelectricity itself [5]. Therefore, NC is of interest not only for applications in electronics, but also for the fundamental understanding of ferroelectric materials themselves.

In the past years, increasing experimental evidence of ferroelectric NC has been reported mainly in well-known peroskite ferroelectrics [6, 7], which appeared to be mainly caused by the movement of ferroelectric domain walls. However, these materials are not suitable for highly scaled transistor applications, due to integration issues and scalability limitations, which is why interest in NC devices initially remained relatively small. This drastically changed after ferroelectricity was surprisingly reported in HfO_2 and ZrO_2 based materials [8, 9], which are already used in current CMOS and DRAM technologies as dielectrics with high permittivity. Ferroelectric properties in these films were observed down to thicknesses below 5 nm, which is promising even for highly scaled CMOS technologies. However, due to the novelty of HfO_2 and ZrO_2 based ferroelectrics, little was known about their potential for NC applications. When my work on the topic of NC began, there was no direct evidence for NC in HfO_2 and ZrO_2 based materials and the basic theoretical understanding of ferroelectric NC in general was very limited, especially regarding the role of domains and domain walls.

Therefore, the focus of this thesis was the theoretical and experimental investigation of ferroelectric NC with a focus on multi-domain effects as well as HfO_2 and ZrO_2 based ferroelectrics, which are most promising for applications in highly integrated circuits. This thesis is structured as follows: In Chapter 2, the fundamental concepts of ferroelectricity and NC are introduced including an overview of the literature on ferroelectric NC and NC devices. In Chapter 3, the experimental fabrication and characterization methods applied in this work are briefly described. Chapter 4 focuses on the experimental observation and modeling of transient NC effects in metal-ferroelectric-metal capacitors. In Chapter 5, the possibility of stabilizing intrinsic NC is theoretically examined, while considering the formation of ferroelectric domains and different electrostatic boundary conditions. Chapter 6 concludes with the experimental observation of stabilized NC under short pulsed voltage operation in ferroelectric/dielectric capacitors, which are proposed to be promising for energy storage applications. Finally, overall conclusions and an outlook are given.

2 Fundamentals

In this chapter, the fundamentals of ferroelectric materials as well as NC are reviewed. In particular, focusing on topics related to the emergence of NC in ferroelectric materials, including the Landau theory of ferroelectric phase transitions, depolarization effects, domain formation and ferroelectric switching. Furthermore, since most of the experimental work in this thesis was carried out using either perovskite or HfO_2 based ferroelectrics, the most important properties of these material classes are reviewed.

2.1 Ferroelectricity

A ferroelectric is a material which has a two or more stable electric polarization configurations in the absence of an applied electric field, i.e. spontaneous polarization states, between which one can switch by the application of an electric field [10, 11]. The origin of ferroelectricity lies in the atomic crystal structure of the material, which gives rise to the spontaneous polarization P_S. The electric displacement field in a ferroelectric material can be written as

$$D = \varepsilon_0 E_f + P = \varepsilon_0 E_f + P_b + P_S = \varepsilon_0 E_f + \varepsilon_0 \chi_b E_f + P_S = \varepsilon_0 \varepsilon_b E_f + P_S, \tag{2.1}$$

where ε_0 is the permittivity of free space, E_f is the electric field, P is the total electric polarization, P_b is the background polarization, χ_b is the background susceptibility and $\varepsilon_b = \chi_b + 1$ is the relative background permittivity of the ferroelectric [12].

2.1.1 Crystal structure

Crystalline materials consist of a periodic arrangement of atoms, which form a crystal lattice. The smallest repeating element of such a lattice is called the unit cell, which describes the fundamental structural unit, which by translation can be used to construct the whole lattice. Therefore, the unit cell contains the complete symmetry information of the (infinite) crystal. By symmetry considerations in three dimensions, all possible crystal structures can be classified into 7 crystal systems, which can be subdivided into 32 point symmetry groups as shown in Table 2.1, using the international Hermann–Mauguin notation. Furthermore, the 32 point groups can be subdivided into 230 space groups, which are commonly used in materials science to distinguish different crystal structures.[13]

Table 2.1: Occurrence of piezo- and pyroelectricity based on the seven crystal systems and 32 point symmetry groups. Only non-centrosymmetric point groups can exhibit piezo- and pyroelectricity.

Crystal system	Point group	Non-centrosymmetric	Piezoelectric	Pyroelectric
triclinic	1	x	x	x
	$\bar{1}$			
monoclinic	2	x	x	x
	m	x	x	x
	$\frac{2}{m}$			
orthorhombic	222	x	x	
	mm2	x	x	x
	$\frac{2}{m}\frac{2}{m}\frac{2}{m}$			
tetragonal	4	x	x	x
	$\bar{4}$	x	x	
	$\frac{4}{m}$			
	422	x	x	
	4mm	x	x	x
	$\bar{4}2m$	x	x	
	$\frac{4}{m}\frac{2}{m}\frac{2}{m}$			
trigonal	3	x	x	x
	$\bar{3}$			
	32	x	x	
	3m	x	x	x
	$\bar{3}\frac{2}{m}$			
hexagonal	6	x	x	x
	$\bar{6}$	x	x	
	$\frac{6}{m}$			
	622	x	x	
	6mm	x	x	x
	$\bar{6}m2$	x	x	
	$\frac{6}{m}\frac{2}{m}\frac{2}{m}$			
cubic	23	x	x	
	$\frac{2}{m}\bar{3}$			
	432	x		
	$\bar{4}3m$	x	x	
	$\frac{4}{m}\bar{3}\frac{2}{m}$			

Out of the 32 point groups, 21 are non-centrosymmetric, meaning that they do not contain an inversion center as one of their symmetry elements. Additionally, 20 out of the 21 non-centrosymmetric point groups are piezoelectric, which means that they exhibit a change of polarization under the application of mechanical strain. Furthermore, half of the piezoelectric point groups possess a spontaneous polarization, which means that they are pyroelectric, since P_S is a function of temperature. The distinction between pyroelectric and ferroelectric crystals is made purely based on the experimental criterion, if the spontaneous polarization can be reversed by the application of an electric field. If the breakdown field strength of the material is lower than the field needed for ferroelectric switching, the material is called pyroelectric, but not ferroelectric. Therefore, it follows that all ferroelectrics are pyroelectric and thus also piezoelectric. This makes ferroelectrics suitable for a wide range of applications also considering their thermal and mechanical properties, e.g. for infrared sensors and ultrasonic transducers.[14, 15]

In a ferroelectric crystal, the microscopic dipole moments of all unit cells align in the same direction. However, in a closely related class of materials called antiferroelectrics, the microscopic dipole moments of neighboring unit cells are anti-parallel, thus resulting in zero macroscopic spontaneous polarization in the absence of an applied electric field [16]. Nevertheless, in such antiferroelectric crystals, the dipoles can be aligned by the application of an electric field, leading to a large macroscopic polarization. Other related materials like ferrielectrics [17] or relaxor-ferroelectrics [18] will not be discussed here.

2.1.2 Landau theory of phase transitions

The Landau theory of phase transitions is the most important phenomenological theory for describing ferroelectricity. Developed in 1937 by Lev Landau [19], this general mean-field theory describes the thermodynamic equilibrium behavior of a substance near a phase transition through the change of an order parameter. Such a phase transition is accompanied by a change in symmetry, where the order parameter is zero in the higher symmetry phase and finite in the lower symmetry phase.[11]

In the case of ferroelectrics, the order parameter is the spontaneous polarization and the phase transition from the low-temperature ferroelectric phase to the high-temperature paraelectric phase happens at the Curie-temperature T_C, where the spontaneous polarization vanishes. Landau's general theory was first applied to ferroelectrics by Ginzburg [20] and Devonshire [21] in 1945 and 1949, respectively, laying the groundwork for what is known today as the Landau-Ginzburg-Devonshire (LGD) theory of ferroelectric phase transitions.[10, 11, 22]

The essence of the theory can be described as follows: The Gibb's free energy density of a ferroelectric can be written as

$$G = \alpha P_S^2 + \beta P_S^4 + \gamma P_S^6 - E_f P_S - \frac{1}{2}\varepsilon_0\varepsilon_b E_f^2 + k\left(\nabla P_S\right)^2, \tag{2.2}$$

where α, β and γ are the material-specific Landau coefficients, k is the domain wall coupling factor and E_f is the electric field inside the ferroelectric [11, 23]. Effects of mechanical strain are neglected in Eq. 2.2. For simplicity, in the following discussion it is assumed that $\varepsilon_b = 0$ and $k = 0$, which assumes that the electrostatic self-energy $\varepsilon_0 \varepsilon_b E_f^2 / 2$ and the domain wall energy $k(\nabla P_S)^2$ are negligible. In this case, the spontaneous polarization P_S is equal to the total polarization P. By minimizing G in Eq. 2.2 with respect to P, one then obtains

$$\frac{dG}{dP} = 0 = 2\alpha P + 4\beta P^3 + 6\gamma P^5 - E_f, \tag{2.3}$$

which results in the ferroelectric equation of state given by

$$E_f = 2\alpha P + 4\beta P^3 + 6\gamma P^5. \tag{2.4}$$

With the temperature dependence of $\alpha = \alpha_0 (T - T_0)$, Eq. 2.2 can then be written as

$$G = \alpha_0 (T - T_0) P^2 + \beta P^4 + \gamma P^6 - E_f P, \tag{2.5}$$

where α_0 is a positive constant and T_0 is the Curie-Weiss temperature (which might not coincide with the Curie-temperature T_C). For $T < T_0$, the parameter $\alpha < 0$.

To discuss the properties of ferroelectric phase transitions, two different cases have to be considered. First, if $\beta < 0$ and $\gamma > 0$ the transition is said to be of first-order. Secondly, if $\beta > 0$ the ferroelectric exhibits a second-order phase transition. In the latter case, the parameter $\gamma \geq 0$.[10, 11]

2.1.2.1 Second-order phase transition

Let us first consider the simpler case of a second-order phase transition, where $\beta > 0$ and $\gamma = 0$. Fig. 2.1a) shows the polarization-free energy landscape without applied electric field for different temperatures relative to the transition temperature T_0. For $T < T_0$, the material exhibits a double-well free energy landscape, which gives rise to two degenerate spontaneous polarization states at zero electric field with magnitude

$$P_0 = \sqrt{\frac{-\alpha_0 (T - T_0)}{2\beta}}. \tag{2.6}$$

This $P_0 > 0$ can also be seen from the polarization-electric field curves in Fig. 2.1b), which has an 'S'-like shape. With increasing temperature, P_0 decreases until it continuously approaches zero at $T = T_0$. Since the spontaneous polarization vanishes at T_0 (see Fig. 2.1c)), this coincides with the

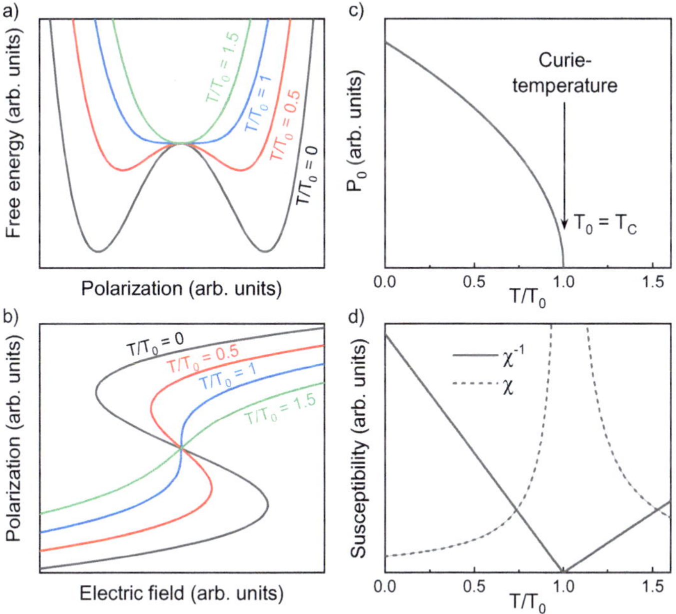

Figure 2.1: Characteristics of a ferroelectric with a second-order phase transition according to Landau theory. a) Free energy-polarization landscapes, b) polarization-electric field curves, c) spontaneous polarization P_0 and d) electric susceptibility χ as a function of the temperature T. T_0 and T_C are the Curie-Weiss temperature and Curie-temperature, respectively.

Curie-temperature T_C of the ferroelectric. Therefore, in a ferroelectric with a second-order phase transition, $T_0 = T_C$. Above T_0, the free energy landscape exhibits only one minimum at $P = 0$, which results in a monotonically increasing relationship between P and E_f. The material is said to be paraelectric for $T > T_0 = T_C$. The electric susceptibility χ of the material at zero electric field in thermodynamic equilibrium is defined as

$$\left.\frac{1}{\chi}\right|_{T<T_C, E_f=0} = \left.\frac{\mathrm{d}^2 G}{\mathrm{d}P^2}\right|_{T<T_C, E_f=0} = \left.\frac{\mathrm{d}E_f}{\mathrm{d}P}\right|_{T<T_C, E_f=0} = 4\alpha_0(T_C - T) \tag{2.7}$$

and

$$\left.\frac{1}{\chi}\right|_{T>T_C, E_f=0} = \left.\frac{\mathrm{d}^2 G}{\mathrm{d}P^2}\right|_{T>T_C, E_f=0} = \left.\frac{\mathrm{d}E_f}{\mathrm{d}P}\right|_{T>T_C, E_f=0} = 2\alpha_0(T - T_C) \tag{2.8}$$

for temperatures below and above T_C, respectively, which is plotted in Fig. 2.1d). The electric susceptibility thus diverges to infinity at $T = T_C$ in a second-order ferroelectric phase transition. Note that so far, only the thermodynamically stable polarization states were discussed, which always results in a positive susceptibility. It will be shown later, that thermodynamically unstable states result in a negative susceptibility and thus cause NC.

2.1.2.2 First-order phase transition

Let us now consider a first-order ferroelectric phase transition with $\beta < 0$ and $\gamma > 0$. In this case, the characteristics of the transition differ strongly from the previously discussed transition of second-order. Depending on the temperature, the free energy landscapes can now also exhibit three energy minima and meta-stable spontaneous polarization states. Indeed, instead of one characteristic temperature T_0 as in the second-order transition, there are three additional characteristic temperatures [10]. First, the Curie-temperature is now defined as the temperature where the non-zero spontaneous polarization ceases to be the lowest energy state, which happens at

$$T_C = T_0 + \frac{\beta^2}{4\alpha_0\gamma}. \tag{2.9}$$

However, above T_C, meta-stable spontaneous polarization states can still exist up until the limit-temperature of the ferroelectric

$$T_1 = T_0 + \frac{\beta^2}{3\alpha_0\gamma}. \tag{2.10}$$

Above T_1, the material can still exhibit a field-induced transition to the ferroelectric state up until

$$T_2 = T_0 + \frac{3\beta^2}{5\alpha_0\gamma}, \tag{2.11}$$

which is the limit-temperature of field-induced ferroelectricity [24]. The temperature T_0 for a first-order transition is called the Curie-Weiss temperature. Fig. 2.2 shows the free energy, polarization-electric field, temperature and susceptibility characteristics of an exemplary first-order ferroelectric phase transition.

The polarization-free energy landscapes in Fig. 2.2a) show that for $T \leq T_0$ only two degenerate energy minima at non-zero polarization exist, similar to the second-order case. However, for $T_0 < T < T_C$, a meta-stable energy minimum at zero polarization emerges, which means that it is higher in energy than the other two minima at finite polarization values. At $T = T_C$, all three minima are degenerate and for $T_C < T < T_1$, the non-zero polarization states become meta-stable.

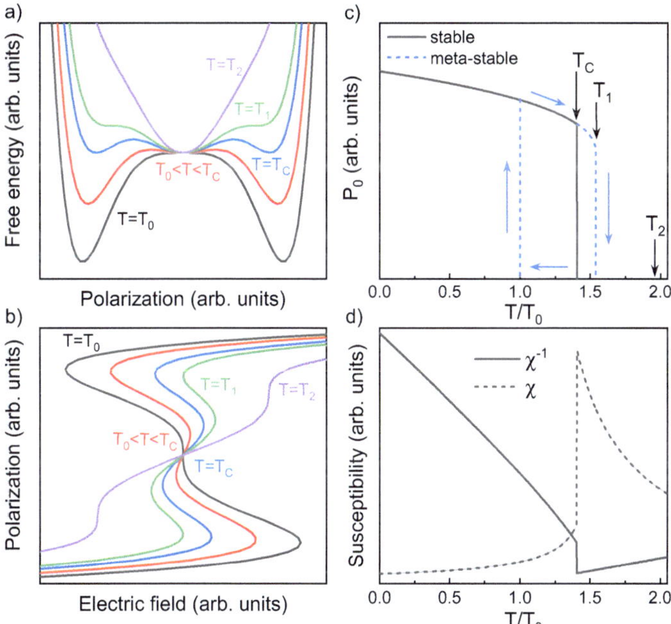

Figure 2.2: Characteristics of a ferroelectric with a first-order phase transition according to Landau theory. a) Free energy-polarization landscapes, b) polarization-electric field curves, c) spontaneous polarization P_0 and d) electric susceptibility χ as a function of the temperature T. T_0, T_1, T_2 and T_C are the Curie-Weiss temperature, the limit-temperature of ferroelectricity, the limit-temperature of field-induced ferroelectricity and the Curie-temperature, respectively.

Furthermore, for $T_1 < T < T_2$, the energy minima at non-zero polarization vanish, such that only at non-zero applied field, a spontaneous polarization is induced. This can also be seen from Fig. 2.2b), where the P-E_f curve has a double-'S'-shaped form. Only for $T > T_2$, the material is completely paraelectric even when an electric field is applied.

The spontaneous polarization states $\pm P_0$ at zero field for a first-order transition have a more complicated dependence on temperature due to the existence of meta-stable states. This is shown in Fig. 2.2c), where the solid and dashed lines correspond to stable and meta-stable regions, respectively. When considering only the stable lowest free energy states, P_0 vanishes at $T = T_C$. However, when meta-stable states are also considered, depending on the history and direction of the temperature change, P_0 exhibits discontinuous jumps at either at $T = T_0$ or $T = T_1$. The relative distance the characteristic temperatures T_0, T_C, T_1 and T_2 depends upon the material specific factor $\beta^2/(\alpha_0\gamma)$.

Now examining the electric susceptibility χ only for the thermodynamically stable state as a function of temperature in Fig. 2.2d), instead of a divergence to infinity as in the second-order transition, a discontinuity of χ can be seen at $T = T_C$. Since both P_0 and χ exhibit such a discontinuity at T_C, the transition is often called a discontinuous first-order transition in contrast to the continuous second-order transition discussed before.

2.1.3 Depolarization

Experimental spontaneous polarization values of ferroelectrics are found to be in the range of around 5 µC cm^{-2} to over 150 µC cm^{-2} [25]. These values are considerably higher compared to electrically induced polarization charges in most dielectric materials. Indeed, without proper external charge screening, the bound polarization charges can cause strong electric fields, which point in the opposite direction of the spontaneous polarization and thus are called depolarization fields. These internal electric fields can have a strong impact on the ferroelectric properties and are closely related to the topic of NC, as will be shown in Section 2.2.

Consider a single crystal ferroelectric with spontaneous polarization $P_S > 0$ in vacuum. Here, the relative background permittivity $\varepsilon_b \neq 0$ is again considered. Assuming that the ferroelectric is infinite in the directions perpendicular to the polarization direction, and assuming that there are no free charges and there is no external electric field, the displacement field D must be zero. Therefore, using Eq. 2.1 the displacement in the ferroelectric can be written as

$$D = 0 = \varepsilon_0 \varepsilon_b E_f + P_S, \tag{2.12}$$

which leads to a depolarization field

$$E_{dep} \equiv E_f = -\frac{P_S}{\varepsilon_0 \varepsilon_b}. \tag{2.13}$$

Note the minus sign in Eq. 2.13, which indicates that the field is anti-parallel to P_S. The depolarization energy density is now given from the expression [23]

$$w_{dep} = -E_{dep}P_S - \frac{1}{2}\varepsilon_0\varepsilon_b E_{dep}^2 = \frac{P_S^2}{2\varepsilon_0\varepsilon_b}. \tag{2.14}$$

When examining Eq. 2.14, it is clear that a large unscreened P_S results in a large depolarization energy, which is unfavorable for the total energy of the ferroelectric. However, the picture changes when a metal-ferroelectric-metal capacitor with shorted electrodes is considered. In this case, assuming idealized ferroelectric/metal interfaces, the free electrons in the metal can perfectly compensate the bound polarization charge, thus reducing E_{dep} to zero. This is shown schematically in Fig. 2.3a), where the color red indicates the bound spontaneous polarization charges and blue corresponds to free charges in the metal electrodes, ρ_{vol} is the charge density per volume, φ is the electrostatic potential and $E = -\nabla\varphi$ is the electric field.

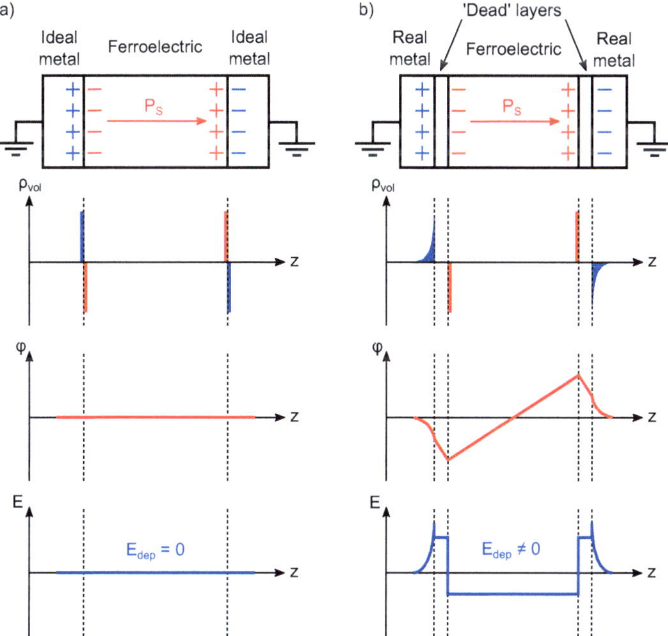

Figure 2.3: Spatial distribution of the charge density ρ_{vol}, the electrostatic potential φ and the electric field E in a) an ideal metal-ferroelectric-metal capacitor and b) a more realistic capacitor with a finite charge screening length in the metal and non-ferroelectric 'dead' interface layers. P_S is the spontaneous polarization and E_{dep} is the depolarization field. Bound polarization charges are red and free metal screening charges are blue.

However, in real metal-ferroelectric-metal (MFM) capacitors, there are two non-ideal effects which lead to finite depolarization fields: First, the finite Thomas-Fermi screening length in the metal [26, 27, 28] and second, non-ferroelectric 'dead' layers at the ferroelectric/metal interfaces [29, 30, 31]. Semiconducting electrodes have an even larger screening length than metals and therefore result in larger depolarization fields [26, 32]. Fig. 2.3b) schematically shows the one-dimensional electrostatics in a more realistic capacitor with both non-ferroelectric interface layers and finite metal screening length. As can be seen, the screening charge in the electrodes is spread out in the z-direction, which leads to a voltage drop inside the metal and thus a depolarization field in the ferroelectric [27]. Interestingly, the presence of interfacial non-ferroelectric (i.e. dielectric) layers has the same qualitative effect as the finite metal screening length [28]. By assuming two symmetric dielectric interface layers of thickness $t_d/2$ and relative permittivity ε_d, the resulting depolarization field in the ferroelectric (assuming ideal metal electrodes) can be written as

$$E_{dep} = -\frac{P_S}{\varepsilon_0 \left(\varepsilon_b + \varepsilon_d \frac{t_f}{t_d}\right)},\tag{2.15}$$

where t_f is the thickness of the ferroelectric. From Eq. 2.15 it can be seen that with decreasing ferroelectric thickness t_f and with increasing dielectric thickness t_d, E_{dep} strongly increases. Furthermore, the lower the relative permittivity of the dielectric layer ε_d, the larger the depolarization. The effect of the finite metal screening length can be understood in a similar way qualitatively, where t_d in Eq. 2.15 would be proportional to the screening length and ε_d would be the permittivity of the metal electrode [28, 33]. While the screening length of typical metals is very small (less than 1 Å), it can still play a significant role in the behavior of ultra-thin ferroelectric films [28].

As a result of the finite depolarization field in any real ferroelectric, several mechanism can result in a reduction of the depolarization energy in the ferroelectric [34], some of which are sketched in Fig. 2.4: Homogeneous depolarization, domain formation or compensation by free charges.

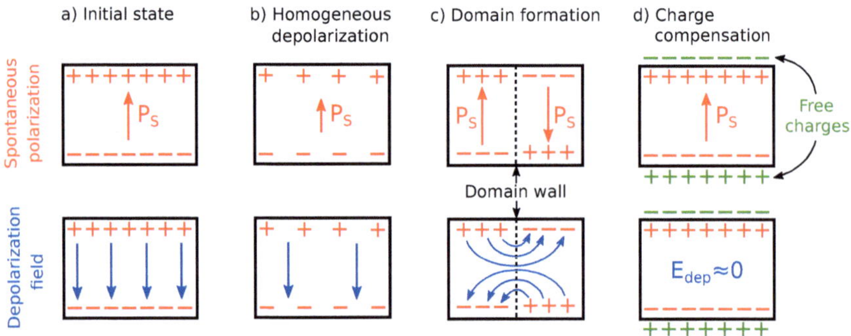

Figure 2.4: Schematic spontaneous polarization P_S and depolarization field E_{dep} configurations for different mechanisms which can reduce the depolarization energy from the initial ferroelectric state in a). b) Homogeneous depolarization, c) domain formation and d) compensation of bound polarization charge (red) by free charges (green). Adapted from Ref. [35].

In the previous discussions, it was assumed that the ferroelectric has a homogeneous polarization and is perfectly insulating. Under these assumptions, a large depolarization field would lead to a homogeneous reduction of P_S, until the total free energy (i.e. the sum of anisotropy energy and depolarization energy) is minimized, see Fig. 2.4b). In this case the depolarization energy is reduced at a cost of an increase in the ferroelectric anisotropy energy.

However, such a homogeneous depolarization is rarely observed in real ferroelectrics, where the spontaneous polarization can be non-uniform [36]. In the simplest case, two homogeneous polarization regions (i.e. domains) of anti-parallel alignment can form, which are separated by a domain wall, as indicated in Fig. 2.4c). In this case, the depolarization fields in polarization direction are strongly reduced at the cost of the domain wall energy. Most of the time, the formation of domains is energetically more favorable than a homogeneous polarization [34, 36, 37]. In real ferroelectrics, much more complicated domain structures can form, e.g. ferroelastic domains with 90° domain walls [38], vortex domains [39] or even skyrmionic domains [40]. However, the investigation of such complicated domain topologies is beyond the scope of this thesis. Interestingly, the movement of ferroelectric domain walls can contribute strongly to the permittivity of the material [41, 42]. These domain wall contributions to the permittivity are often called "extrinsic" in contrast to the "intrinsic" permittivity of the volume of a ferroelectric domain [43, 44].

Lastly, since ferroelectric and also potential non-ferroelectric interface layers have finite conductivity [10, 34], free charges might be able to move through the layers to compensate the bound polarization charge, see Fig. 2.4d). There are two potential origins of such free charges. First, for lower bandgap ferroelectrics and ferroelectrics with defect bands close to the band edges, finite conductivity of the ferroelectric itself can lead to internal compensation of the bound charge [45]. On the other hand, since the non-ferroelectric interface layers are typically very thin (less than 2 nm), electrons can tunnel through these layers from the electrodes and might get trapped at interface defects, thus compensating part of the spontaneous polarization. Especially in polycrystalline ferroelectric films, significant amounts of defects at grain boundaries and interfaces might favor charge compensation mechanisms for the reduction of the depolarization energy. Another possibility to reduce E_{dep}, which is not shown in Fig. 2.4 is the reorientation of P_S from the out-of-plane to the in-plane direction [34], which is not discussed here.

2.1.4 Ferroelectric switching

So far, mainly the thermodynamic equilibrium conditions and basic electrostatics of ferroelectrics have been described. However, for practical applications, the ferroelectric polarization dynamics, i.e. switching kinetics are of critical interest. Especially in the field of NC, this is one of the central questions which have to be answered to assess if NC can be used in high-speed applications in electronics [46, 47, 48, 49, 50, 51]. Experimentally, when an electric field is applied to a dielectric capacitor, and the flowing charge (i.e. the polarization) is measured, a linear relationship between P and E is observed as shown in Fig. 2.5a). In contrast, if large enough electric fields are applied to a ferroelectric capacitor, a distinct hysteresis-loop is obtained in the P-E_f characteristics, see Fig. 2.5b).[10]

The polarization which remains at zero electric field is called the remanent polarization P_r, which in an ideal ferroelectric capacitor with out-of-plane polarization direction would correspond to the spontaneous polarization P_S. However, many non-ideal effects can lead to a significant difference between the theoretical P_S and the experimental P_r, such as crystal orientation, morphology, non-ferroelectric phase fractions, defects and leakage currents [52]. The electric field at which the polarization changes sign is called the coercive field E_c. The experimental value of E_c is of critical significance when investigating ferroelectric switching dynamics. In principle, two different

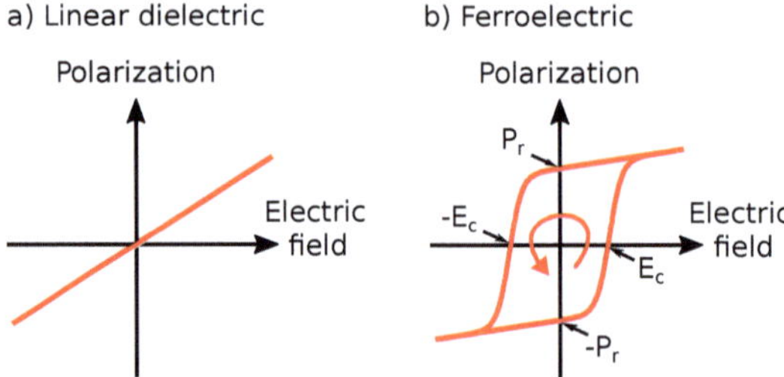

Figure 2.5: Schematic polarization-electric field dependencies of a) a linear dielectric and b) a ferroelectric material. P_r and E_c are the remanent polarization and the coercive field, respectively. Adapted from Ref. [35].

ferroelectric switching effects can be distinguished: Intrinsic and extrinsic (domain-mediated) ones.[53, 54, 55, 56]

The intrinsic, or thermodynamic switching case can be understood from the previously discussed homogeneous Landau approach [53, 56]. In this case, the intrinsic E_c is related to the barrier height of the free energy landscape. Therefore, as can be seen e.g. in Fig. 2.1b), the thermodynamic coercive field is given by the points of infinite slope dP/dE_f of the 'S'-shaped P-E_f curve. Since a homogeneous polarization was assumed in this simplified model, the theoretical intrinsic coercive field is typically orders of magnitude larger than the experimental one. Nevertheless, some reports have suggested that in some very thin ferroelectric layers, intrinsic switching can occur [53, 56].

The extrinsic switching case is mainly governed by reverse domain nucleation and growth, which, depending on the film properties, can be either limited by the nucleation process or the speed of domain wall motion [57]. While the latter case is well described by the so-called Kolmogorov-Avrami-Ishibashi model [58], the former behavior is often associated with the Du-Chen model [59] or nucleation-limited switching model [60]. In any case, extrinsic domain-mediated switching, due to strongly inhomogeneous polarization dynamics, should have completely different experimental characteristics compared to intrinsic switching. For example, the energy barrier for domain nucleation and growth was shown to be orders of magnitude smaller than the thermodynamic barrier, which also results in a much lower experimental coercive field for extrinsic switching dynamics [54].

2.1.5 Material classes

Ferroelectricity was first discovered in 1920 by Valasek in crystals of Rochelle Salt [61], which is known today to be one of the most complicated ferroelectric materials from the crystallographic point of view [62]. In the 1930s, ferroelectric properties at low temperatures were discovered in potassium dihydrogen phosphate [63]. However, the research interest in ferroelectrics significantly increased with the discovery of ferroelectricity in materials with perovskite structure (i.e. similar to $CaTiO_3$), the first of which was $BaTiO_3$ [64]. Other classes of ferroelectrics [14], such as polymer ferroelectrics (e.g. poly[vinylidenefluoride-co-trifluoroethylene], P(VDF-TrFE)) [65] or Triglycine Sulphate based materials [66], will not be discussed here.

2.1.5.1 Perovskite ferroelectrics

Besides $BaTiO_3$, one of the most popular ferroelectric materials systems since the 1950s was the solid solution of lead titanate ($PbTiO_3$) and lead zirconate ($PbZrO_3$), which is called lead zirconate titanate ($Pb(Zr_{1-r}Ti_r)O_3$ or PZT), where $0 < r < 1$ [67, 68]. The simplified phase diagram of PZT is shown in Fig. 2.6a). While pure $PbZrO_3$ was found to be antiferroelectric at room temperature [69], higher fractions of Ti result in the formation of rhombohedral and tetragonal ferroelectric phases. The Curie-temperature T_C generally increases with increasing $PbTiO_3$ content. The ferroelectric PZT layers used in this work had a $PbZr_{0.2}Ti_{0.8}O_3$ composition and thus were in the ferroelectric tetragonal phase, far from the morphotropic phase boundary between the rhombohedral and tetragonal ferroelectric phases [70].

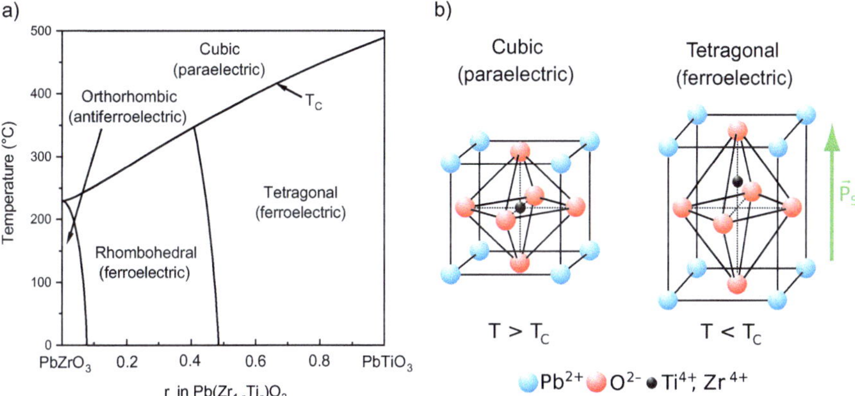

Figure 2.6: a) Phase diagram for the solid solution of lead zirconate and lead titanate ($Pb(Zr_{1-r}Ti_r)O_3$) as a function of temperature, where T_C is the Curie-temperature (after Ref. [70]). b) Schematic unit cells for the paraelectric cubic and ferroelectric tetragonal phase. P_S is the spontaneous polarization (adapted from https://commons.wikimedia.org/wiki/File:Perovskite.svg).

The schematic unit cells for the paraelectric cubic and the ferroelectric tetragonal phase of PZT are shown in Fig. 2.6b). The vertices of the cubic unit cell consist of Pb atoms, while the central Ti or Zr atom is surrounded by an oxygen octahedron. In the tetragonal phase, the Ti/Zr atom shifts from the center into one of the two symmetric off-center positions, which gives rise to the spontaneous polarization, see Fig. 2.6b) [71].

For non-volatile memory applications, the layered perovskite strontium bismuth tantalate was later also applied [72, 73]. However, all of these perovskite ferroelectrics like $BaTiO_3$, PZT and strontium bismuth tantalate have several drawbacks for applications in integrated circuits: They are not compatible with CMOS technology, have a relatively low bandgap, degrade under forming gas annealing, cannot be deposited in a conformal way on 3D structures and have a limited thickness-scalability [74, 75]. Due to these reasons, non-volatile memories based on ferroelectric perovskites have been limited to niche applications, where low power, fast write speed and high cycling endurance is needed [76]. However, due to the aforementioned problems, commercial ferroelectric random access memory has not been able to compete with the much cheaper, higher memory density alternatives, like NAND-Flash so far.

2.1.5.2 Fluorite-type ferroelectrics

In 2011, surprisingly, Böscke et al. reported ferroelectric and antiferroelectric properties in thin layers ($\sim$10 nm) of Si-doped HfO_2 [8]. This was unexpected, since at that time HfO_2 was already used as the standard high-permittivity gate dielectric in advanced CMOS technology for many years. Furthermore, ferroelectric properties were also found for many other dopants in HfO_2 [77, 78, 79, 80, 81] and in the solid solution of HfO_2 and ZrO_2, where an equal mixture resulted in the strongest ferroelectric response, while pure ZrO_2 was found to be antiferroelectric [82, 9, 83]. Even undoped HfO_2 based thin films were shown to exhibit ferroelectricity [84, 85]. In analogy to perovskite ferroelectrics, these materials have been recently called fluorite-type ferroelectrics [86] due to their crystallographic similarities with calcium fluoride CaF_2.

The known bulk phases of HfO_2 and ZrO_2 at atmospheric pressure are the monoclinic $P2_1/c$ phase at room temperature as well as the tetragonal $P4_2/nmc$ and the cubic $Fm\bar{3}m$ phases at higher temperatures [8]. However, it was shown that in thin films in the range below 10 nm, the higher symmetry tetragonal and cubic phases can be stabilized even at room temperature due to the surface energy effect and small grain sizes [87]. The addition of doping and capping with metallic layers were further shown to enhance the formation of these phases especially in HfO_2, which was of great commercial interest to increase the relative permittivity of dielectric layers in CMOS logic transistors and DRAM further [88]. With this goal in mind, the unexpected emergence of a non-centrosymmetric orthorhombic phase of space group $Pca2_1$ was discovered to result in ferroelectric behavior [8]. This orthorhombic phase was first experimentally observed in Mg stabilized ZrO_2 particles in a cubic matrix [89].

The unit cells of the tetragonal and orthorhombic phase with the two opposite polarization directions are shown in Fig. 2.7. The original understanding of the stabilization of the ferroelectric orthorhombic phase in capacitors capped with TiN is also indicated in this figure: First, at elevated temperature, the HfO_2 thin film nucleates in the tetragonal phase and during cooling transforms into the orthorhombic phase due to the mechanical confinement from the top electrode [8, 90].

Furthermore, it was found that also the doping concentration, film thickness, thermal budget and presence of oxygen vacancies can strongly influence the formation of the orthorhombic phase in polycrystalline HfO_2 thin films [91]. In such polycrystalline films, the ferroelectric phase transitions are typically much broader compared to the theoretical ones based on Landau theory [90], which can be understood based on the inhomogeneous film properties, e.g. the distribution of grain sizes and orientations [92, 93, 94].

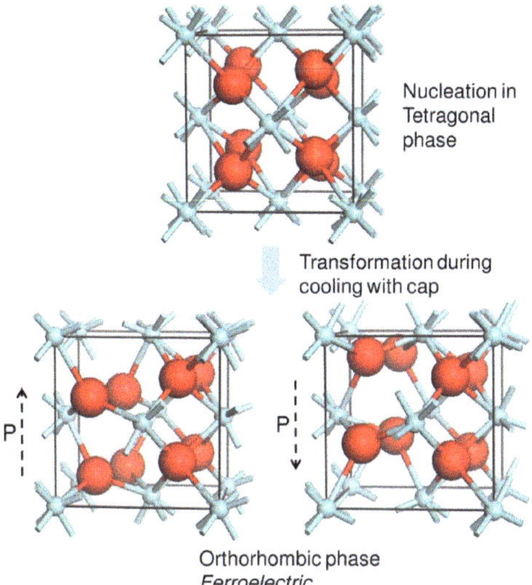

Figure 2.7: Hafnium oxide (HfO_2) unit cells for the paraelectric tetragonal and the ferroelectric orthorhombic phases and the schematic mechanism for the proposed stabilization procedure. Reprinted from [8], with the permission of AIP Publishing.

Interestingly, antiferroelectric hysteresis loops observed in HfO_2 and ZrO_2 based thin films were suggested to result from an electric field-induced phase transition from the tetragonal $P4_2/nmc$ phase to the orthorhombic $Pca2_1$ phase [8, 95, 96], which is in contrast to other antiferroelectric materials like $PbZrO_3$, where even the ground state has a anti-parallel aligned spontaneous polarization in the unit cell [69].

Compared to perovskite ferroelectrics, the coercive field for fluorite-type ferroelectrics is typically much larger in the range of 1–2 MV cm^{-1} [97]. The theoretical spontaneous polarization of the orthorhombic $Pca2_1$ phase has been calculated from first principles to be between 50–60 μC cm^{-2} [98, 99, 100, 101], which is comparable to the values known from perovskite ferroelectrics. Experimental remanent polarization values are naturally lower due to the distribution of grain orientations and non-ferroelectric phase fractions in the films, leading to 10–30 μC cm^{-2} as typical values [97, 91]. Due to their polycrystalline structure and small thickness, depolarization effects play an important role in their ferroelectric properties [102, 103, 104]. Furthermore, the relative permittivity of ferroelectric or antiferroelectric HfO_2 and ZrO_2 based thin films is typically in the range from 25 up to 55 [97, 105], which is much lower compared to typical perovskite ferroelectrics, which have typical relative permittivities of several hundreds up to thousands [14, 106]. The lower permittivity of fluorite-type ferroelectrics has advantages for the integration into ferroelectric memories [82] and furthermore is related to a higher electronic band gap and therefore a higher breakdown field strength [107].

2.2 Negative capacitance

When the electrostatic potential φ of a conductor changes by an infinitesimal amount $d\varphi$, the charge on the conductor Q changes by dQ. Both quantities are related by the proportionality constant C_{self} as

$$dQ = C_{self}d\varphi. \tag{2.16}$$

In the case of a single conductor of charge Q and potential φ, C_{self} is called the *self-capacitance*. However, what is typically meant when the term capacitance is used, is the *mutual capacitance* between two different conductors, with the potentials φ_1 and φ_2, where the voltage between them is defined as $V = \varphi_1 - \varphi_2$. In practice, conductor 1 has the positive charge $+Q$ and conductor 2 has the opposite charge $-Q$. In this case, a small change in voltage dV leads to a change in charge dQ on both conductors as

$$dQ = CdV, \tag{2.17}$$

where C is the (mutual) capacitance, which is generally defined as

$$C = \frac{dQ}{dV}. \tag{2.18}$$

In the case of ideal static conductors surrounded by vacuum or any ideal linear dielectric materials, C is a constant independent of Q and V. Thus it can be written

$$C = \frac{Q}{V}. \tag{2.19}$$

In this case, C is dependent only on the geometry of the system as well as the permittivities of the linear dielectric materials. In reality, there are many effects which can lead to a non-linear capacitance behavior, e.g. non-linear polarizable materials or the finite conductivity of insulators. Therefore, Eq. 2.19 is only valid as a simplified approximation, but the more general case is described by Eq. 2.18. Nevertheless, it turns out that Eq. 2.19 is a very good approximation in many practical situations. For the case of ferroelectrics and NC on the other hand, using Eq. 2.18 is mandatory.

For a regular parallel plate capacitor with a plate area A, distance between the electrodes d_c and relative permittivity of the dielectric material in between ε_d, the capacitance when neglecting fringing fields is simply given by

$$C = \frac{\varepsilon_0 \varepsilon_d A}{d_c}. \tag{2.20}$$

Since generally A and d_c are positive quantities and $\varepsilon_0 = 8.854 \cdot 10^{-12} \, \mathrm{As\,V^{-1}\,m^{-1}}$, the phenomenon of NC is closely related to the question of negative permittivity.

2.2.1 Can capacitance be negative?

In analogy to negative resistance circuit elements, in the 1930s it was proposed that circuit elements with NC or negative inductance could also be utilized [108]. While NC can be relatively easily achieved in active circuits, such as in a negative impedance converter using operational amplifiers [109], the focus of this thesis is on (differential) NC in passive media.

In general, there are many different physical effects which can give rise to NC. For example, in the optical frequency range, metals exhibit a negative permittivity below the plasma frequency, which typically lies in the ultra-violet electromagnetic spectrum [110, 111]. This effect is utilized in applications in plasmonics and in the field of meta-materials [112, 113, 114]. Typically, at much lower frequencies, NC has been observed in semiconductor structures like infrared detectors, solar cells, light emitting diodes and metal-semiconductor interfaces, among others [115, 116, 117, 118, 119]. The origin of these NC effects have been explained by Jonscher by considering the time domain [120]. He argued, that the current response to a step voltage function of a NC is characterized by a delayed increase in current, which might occur due to different physical mechanisms. However, Jonscher only considered NC by itself, which in passive media can only be observed temporarily, since it is unstable by itself [120]. Furthermore, nanoelectromechanical switches show regions of NC since the distance between the electrodes changes with the applied voltage [121, 122, 123, 124].

The original idea that the NC in a passive device could be stabilized by an external circuit can be attributed to Rolf Landauer [1]. Using the example of a ferroelectric material, he argued that under certain conditions (e.g. domains cannot form), putting a negative resistance in parallel to the ferroelectric should in principle result in a stabilization of the NC. He was also the first to predict that transient NC might be observed in ferroelectric capacitors during switching [1].

2.2.2 Ferroelectric negative capacitance

For a long time after Landauer's original paper, there was no research on NC in ferroelectrics. Indeed, the next investigation of NC in ferroelectrics with a periodic domain structure was published in 2001 by Bratkovsky and Levanyuk [42], who later showed that the P-E_f hysteresis loop in ultrathin ferroelectric layers (when considering the depolarization field) can have a negative slope [125]. However, in both cases the NC was found to be a result of the sideways motion of 180° domain walls. In contrast, the original suggestion of Salahuddin and Datta in 2008 of an NC transistor was based on a simplified single-domain Landau model [4]. However, it was argued that any mechanism which involves a similar microscopic positive feedback mechanism could result in NC, including inhomogeneous domain switching [4].

In hindsight, many of the misunderstandings about NC might be traced back to the application of the single-domain Landau model in the original publication by Salahuddin and Datta, which can give qualitatively different results than other multi-domain models in many situations. Therefore, it is important to first understand the differences between these single-domain (intrinsic) and multi-domain (extrinsic) NC effects.

2.2.2.1 Intrinsic single-domain negative capacitance

Intrinsic NC, which has also been called incipient NC [126], is based on the idea of a homogeneously depolarized ferroelectric material. Using the homogeneous Landau theory introduced in Eq. 2.4 and 2.5 for a second-order phase transition, the NC region can be identified as the green regions in Fig. 2.8. In the double-well Helmholtz free energy landscape ($F_f = G + E_f P$) in Fig. 2.8a), according to Eq. 2.7 the inverse ferroelectric susceptibility is proportional to the curvature of the energy landscape [127]. This means that the NC region corresponds to the energy barrier between the two energy minima, where the ferroelectric by itself is thermodynamically unstable [10, 26].

In the P-E_f curve in Fig. 2.8b), the susceptibility is proportional to the slope $\mathrm{d}P/\mathrm{d}E_f$, i.e. NC is expected at the regions of negative slope in any P-E_f curve. This would still hold true for a first-order phase transition, where even two separate regions of negative slope might occur, see Fig. 2.2b). In the case of this homogeneous model, NC is a direct consequence of the thermodynamic polarization instability in the ferroelectric [127]. The dynamic behavior of this intrinsic NC can be modeled with the Landau-Khalatnikov equation [4]

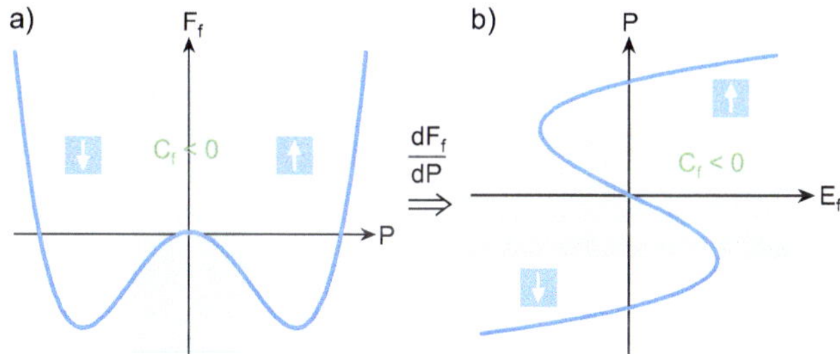

Figure 2.8: The ferroelectric NC ($C_f < 0$) region as a function of polarization in Landau theory. a) Helmholtz free energy F_f-polarization P landscape and b) polarization-electric field E_f dependence of a ferroelectric with second-order phase transition. Adapted from Ref. [128].

$$\rho\frac{\partial P}{\partial t} + \frac{\partial G}{\partial P} = 0, \tag{2.21}$$

where ρ is a damping constant of unit [Ωm], which can be interpreted as an internal series resistance. Using Eq. 2.5, this then yields for the field in the ferroelectric

$$E_f = 2\alpha P + 4\beta P^3 + 6\gamma P^5 + \rho\frac{dP}{dt}. \tag{2.22}$$

However, recently it was shown that NC also arises naturally without invoking Landau theory based on the long-range electrostatic interaction of microscopic electric dipoles [5, 129, 130]. In these works it was found that the origin of ferroelectricity itself, namely the polarization catastrophe, which is triggered from a positive feedback loop between the local electric field and the electric dipole moment, leads to a NC which spontaneously polarizes the ferroelectric [5, 129].

2.2.2.2 Extrinsic multi-domain negative capacitance

In contrast to the intrinsic NC discussed so far, extrinsic NC is caused by the movement of ferroelectric domain walls [42, 51, 125, 7]. Since ferroelectric domain wall motion can be completely reversible under certain conditions, extrinsic NC effects can also be hysteresis-free [51]. The simplest multi-domain structure which can exhibit NC is a metal-insulator-ferroelectric-insulator-metal capacitor with two anti-parallel domains as shown Fig. 2.9a) when no voltage is applied [42]. In this case, the domain wall is located at $x = 0$ and both domains have the same size $d/2$, where d is the domain period.

In this simplified model, the spontaneous polarization inside each domain has the constant value P_0 and the domain wall is assumed to be abrupt (Kittel approximation). Due to the symmetry in

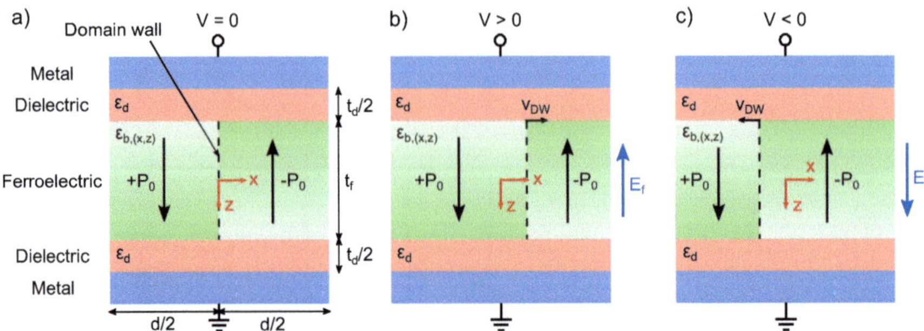

Figure 2.9: NC from a 180° stripe domain pattern in a metal-dielectric-ferroelectric-dielectric-metal capacitor. a) Equilibrium when no voltage is applied ($V = 0$). b) and c) show domain wall displacements for positive and negative applied voltages, respectively. P_0, v_{DW}, ε_d, $\varepsilon_{b,(x,z)}$, t_d, t_f, d and E_f are the spontaneous polarization, domain wall velocity, dielectric permittivity, ferroelectric background permittivity, dielectric thickness, ferroelectric thickness, domain period and electric field in the ferroelectric, respectively. Based on figures in Refs. [51, 131].

Fig. 2.9a), the average polarization and electric fields in the ferroelectric are zero. When a positive voltage is applied, as shown in Fig. 2.9b), the domain wall moves in the positive x-direction with the velocity v_{DW}. After some time the system will reach its equilibrium with the new domain wall position at $x > 0$. Now, the net spontaneous polarization will be positive, leading to an average negative depolarization field in the ferroelectric. Analogously, when applying a negative voltage, the domain wall will be displaced into the negative x-direction, leading to a net negative spontaneous polarization and thus a positive field in the ferroelectric, see Fig. 2.9c).[51, 131]

Microscopically, the redistribution of inhomogeneous stray depolarization fields at the ferroelectric/dielectric interfaces will result in a negative contribution to the ferroelectric permittivity [42, 51, 125, 7]. In the limit where the dielectric thickness is large enough compared to the domain width $d/2$, such that the inhomogeneous stray fields can decay inside the dielectric layers, the total ferroelectric permittivity can be written as [7, 131]

$$\varepsilon_f = \varepsilon_{b,z} - \frac{\pi}{2\ln(2)} \sqrt{\frac{\varepsilon_{b,x}}{\varepsilon_{b,z}}} \frac{t_f}{d} \left(1 + \frac{\varepsilon_d}{\sqrt{\varepsilon_{b,x}\varepsilon_{b,z}}} \right) \varepsilon_{b,z}, \tag{2.23}$$

where $\varepsilon_{b,x}$ and $\varepsilon_{b,z}$ are the in-plane and out-of-plane relative background permittivities of the ferroelectric, respectively. However, Eq. 2.23 shows that the NC term due to domain wall motion dominates if the term t_f/d is large, which is true in the Landau-Kittel limit [131, 7]. However, for thinner dielectric layers, this approximation breaks down. For this more general case, Park et al. developed a dynamic compact model [51]. In this model, the velocity of the ferroelectric domain wall $v_{DW} = dx/dt$ at its position x can then be written as

$$\frac{\mathrm{d}x}{\mathrm{d}t} = \frac{\mu_{DW}}{t_f}\left(\frac{C_d}{C_d + C_{f,b}}V(t) + \frac{1}{C_d + C_{f,b}}\frac{2xP_0}{d} + \frac{2dP_0}{\pi^2\varepsilon_0}\sum_{n=1}^{\infty}\frac{(-1)^n}{n^2 D_n}\sin\left(2n\pi\frac{x}{d}\right)\right),\qquad(2.24)$$

where μ_{DW} is the domain wall mobility, C_d is the dielectric capacitance per area, $C_{f,b}$ is the ferroelectric background capacitance per area and the coefficients D_n are given by

$$D_n = \sqrt{\varepsilon_{b,x}\varepsilon_{b,z}}\coth\left(\sqrt{\frac{\varepsilon_{b,x}}{\varepsilon_{b,z}}}\pi n\frac{t_f}{d}\right) + \varepsilon_d\coth\left(\pi n\frac{t_d}{d}\right).\qquad(2.25)$$

The net spontaneous polarization in the ferroelectric is given by $\langle P_S\rangle = 2xP_0/d$ and the average field field inside the ferroelectric can be written as

$$\langle E_f\rangle = E_{f,ext} + \langle E_{dep,h}\rangle = \frac{C_d V(t) + \langle P_S\rangle}{(C_d + C_{f,b})t_f},\qquad(2.26)$$

where $E_{f,ext}$ and $\langle E_{dep,h}\rangle$ are the homogeneous external and average depolarization field contributions, respectively. Interestingly, when plotting $\langle P_S\rangle$ as a function of $\langle E_f\rangle$, the resulting $\langle P_S\rangle$-$\langle E_f\rangle$ curve has an 'S'-shape, qualitatively similar to the Landau-based homogeneous P_S-E curve in Fig. 2.8b). However, there is an important difference: The shape of the $\langle P_S\rangle$-$\langle E_f\rangle$ 'S'-curve is an extrinsic property of the ferroelectric domain pattern and the electrostatics in the structure. In particular, it depends on the domain period, which is a function of the ferroelectric and dielectric thicknesses t_f and t_d. This is in strong contrast to the intrinsic NC derived from Landau theory, where the 'S'-shape of the intrinsic P_S-E curve is independent of both t_f and t_d.[51]

As is well known from classical ferroelectrics, the equilibrium domain period of a 180° domain pattern is proportional to the square root of the ferroelectric thickness, which is known as the Landau-Lifshitz-Kittel scaling law [132, 133]:

$$d = \sqrt{s_0 t_f},\qquad(2.27)$$

where s_0 is a constant. On the other hand, the domain period d is only a weak function of the dielectric thickness t_d, if t_d is much larger than the domain wall thickness [30]. However, if t_d is comparable to the domain wall thickness, d strongly increases with decreasing t_d [30]. Generally, for very large t_d compared to the equilibrium domain width, d will be independent of t_d, since the inhomogeneous stray electric fields inside the dielectric will decay before reaching the metal electrodes [134, 135]. This means that the stray depolarization energy does not change beyond a certain t_d, making the $\langle P\rangle$-$\langle E_f\rangle$ 'S'-curve independent of t_d in this regime. Nevertheless, for t_d thinner than $d/2$, the $\langle P\rangle$-$\langle E_f\rangle$ curve will get narrower with decreasing t_d due to the partial screening of polarization charge through the electrodes, which changes the energy of the stray depolarization fields [134, 135].

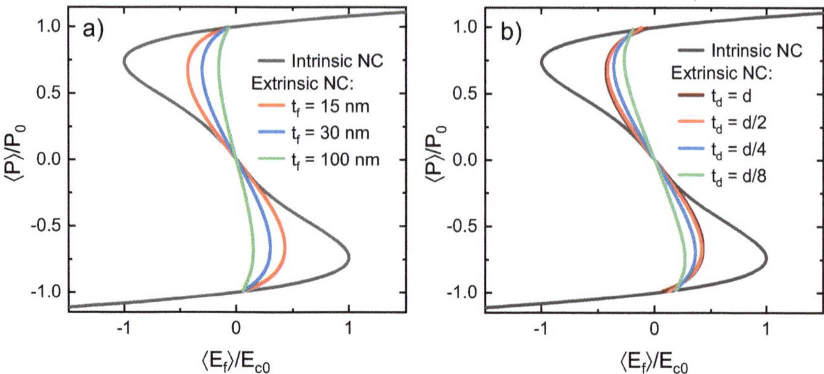

Figure 2.10: Intrinsic and extrinsic average polarization $\langle P \rangle$-electric field $\langle E_f \rangle$ curves normalized to the spontaneous polarization P_0 and intrinsic coercive field E_{c0} for a) different ferroelectric thickness t_f and b) different dielectric thickness t_d with regard to the domain period d. Based on the compact model and modeling parameters from Ref. [51].

Fig. 2.10a) and b) show the $\langle P \rangle$-$\langle E_f \rangle$ curves and the homogeneous Landau P-E_f curve when changing the ferroelectric and dielectric thickness, respectively. Simulation parameters used for solving Eq. 2.24 were $\mu_{DW} = 1.6 \cdot 10^{-6}$ m^2 V^{-1} s^{-1}, $\varepsilon_{b,x} = 417$, $\varepsilon_{b,z} = 174.5$, $\varepsilon_d = 208$, $t_f = 15$ nm, $t_d = 7$ nm, $s_0 = 4.5$ nm and $P_0 = 0.265$ C m^{-2} if not specified otherwise, similar to the values used by Park et al. in Fig. 15.[51]

2.2.2.3 Transient and stabilized negative capacitance

In the previous discussion only the origin of intrinsic and extrinsic NC effects was discussed. However, one important aspect of the idea of NC is stabilization [4], which is closely related to the observation of a polarization hysteresis [136]. However, for most NC applications, hysteresis is undesirable [4, 129]. As briefly mentioned in the section on intrinsic NC, the NC region by itself is thermodynamically unstable. Nevertheless, such an unstable NC might be observed experimentally in a transient measurement, where polarization kinetics dominate [127]. Such hysteretic NC effects have been termed *transient* NC, to distinguish them from stabilized NC without hysteresis [127, 137, 138, 139]. In principle, both intrinsic and extrinsic NC can be either transient or stabilized, leading to four different possible NC phenomena.

From the original paper by Salahuddin and Datta [4], it was proposed that a positive series capacitance can stabilize the intrinsic NC of a ferroelectric. For a second-order phase transition ferroelectric in series with a positive capacitance C_d, the resulting total capacitance is given by

$$\frac{1}{C} = \frac{1}{C_f} + \frac{1}{C_d} = 2\alpha t_f + 12\beta P^2 t_f + \frac{1}{C_d}. \tag{2.28}$$

For the total system to be stable at $P = 0$, the total capacitance must be positive leading to

$$\frac{1}{C} \approx 2\alpha t_f + \frac{1}{C_d} > 0, \tag{2.29}$$

which means that the intrinsic ferroelectric NC should be stable below the critical ferroelectric thickness

$$t_f \leq \frac{1}{2\alpha C_d}. \tag{2.30}$$

As discussed before, Eq. 2.30 is only valid if ferroelectric domains cannot form. The condition of NC stabilization can be visualized using the load-line method, where the positive capacitance C_d corresponds to a load-line in the P-E_f graph, which shifts when a voltage is applied to the series connection of both capacitances, see Fig. 2.11. The intersections mark the operating points of the system. The slope of the load-line depends on the value of C_d.[4]

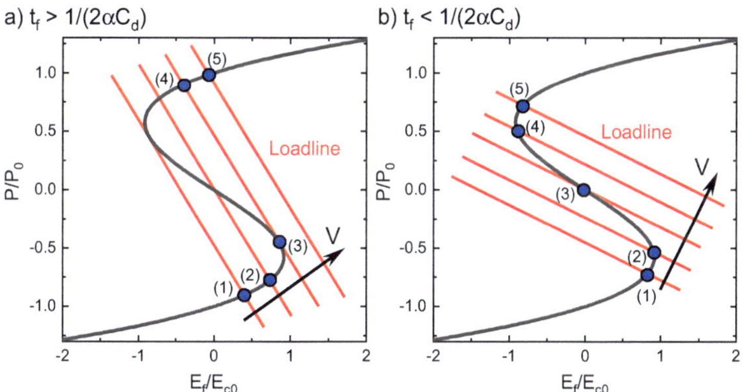

Figure 2.11: Load-line analysis of a second-order intrinsic Landau ferroelectric and a positive capacitance C_d in series a) when the ferroelectric thickness is larger and b) when it is smaller than the critical thickness for stabilization of intrinsic NC. Operating points are marked blue. The load-line (red) shifts when a voltage V is applied. P_0, E_{c0} and α are the spontaneous polarization at zero electric field, the intrinsic coercive field and the first Landau coefficient, respectively. Based on the Supporting Information of Ref. [4].

In the case where Eq. 2.30 is not satisfied, the NC region cannot be traversed without hysteresis as shown in Fig. 2.11a). When starting from operating point (1) and increasing the voltage, the system jumps from point (3) to (4) without going through the complete NC region. When going back, a similar jump would occur towards negative polarization at a different voltage thus leading to hysteresis. However, transient NC might still be observed in this case. On the other hand, if Eq. 2.30 is satisfied, there is always only one intersection of the load-line with the ferroelectric 'S'-curve, as shown in Fig. 2.11b). Therefore, when increasing and decreasing the voltage, the same curve without hysteresis can be traced by the operating points. In this case, the voltage across C_d is larger than the applied voltage, leading to a differential voltage amplification, which would be useful for many applications [4]. Indeed the total capacitance will be larger than C_d, which would be impossible for two positive capacitors in series. The better the capacitances of the positive capacitor and the ferroelectric are matched (i.e. the closer t_f is to the critical thickness in Eq. 2.30), the larger the voltage amplification. For perfect capacitance matching $1/t_f = 2\alpha C_d$, the maximum voltage amplification even becomes infinite.

In the case of extrinsic NC, the situation is different, since the NC depends on the domain structure and dynamics in the ferroelectric. Here, the incomplete screening of the bound polarization charge is responsible for the depolarization which is necessary for the NC effect. When switching a typical metal-ferroelectric-metal capacitor, the change of polarization is governed by domain nucleation and growth dynamics, which necessarily involve hysteresis. However, when the supply of screening charge on the metal electrodes is slowed down, e.g. by the external circuitry, the polarization might change faster than the metal screening, charge leading to a transient NC in such a case.[127, 138]

Nevertheless, as discussed before, reversible domain wall motion in a 180° stripe domain pattern can give rise to hysteresis-free extrinsic NC [42, 51]. The stability of the non-hysteretic region depends on several factors: The layer thicknesses and electrostatics, the possibility of charge compensation by electron tunneling through the dielectric layers and subsequent trapping or de-trapping of mobile charges. Park et al. found that for thicker ferroelectric films and larger applied fields, the polarization dynamics transitioned from a hysteresis-free stripe-domain pattern to a nucleation and growth mechanism with hysteresis [51]. However, the exact necessary boundary conditions for hysteresis-free extrinsic NC are still not completely understood. Recently, it was found that even in complicated vortex-like domain structures in ferroelectric/paraelectric superlattices, NC could be stabilized [140]. In this case, the observed stabilized NC seems to result from a superposition of localized intrinsic and extrinsic NC regions.

2.2.3 Negative capacitance field-effect transistors

While not the focus of this thesis, the original goal of using NC was the application in negative capacitance field-effect transistors (NCFETs) [4]. Therefore, a short overview on the working principle and state-of-the-art of NCFETs will be given here. For a more thorough review of recent works on NCFETs, see e.g. Refs. [124, 141, 142, 143].

The power dissipation in MOSFETs is mainly limited by the magnitude of the power supply voltage V_{dd} for a given ratio between the high drain current on-state ($I_d = I_{on}$ for $V_{gs} = V_{dd}$) and the low drain current off-state ($I_d = I_{off}$ for $V_{gs} = 0$), where I_d is the drain current and V_{gs} is the gate-source voltage. The total power dissipation of a MOSFET is proportional to [144]

$$P_{diss} \propto CV_{dd}^2 f + I_{off}V_{dd}, \tag{2.31}$$

where C is the load capacitance and f is the clock frequency. The most effective measure to reduce P_{diss} is therefore to reduce V_{dd}. Furthermore, reducing V_{dd} is also necessary to keep the electric fields in the device constant during geometrical scaling [145]. However, for a MOSFET with fixed I_{off}, reducing V_{dd} will also reduce I_{on} and thus degrade performance. On the other hand, for a MOSFET with fixed I_{on}, V_{dd} can be reduced by reducing the threshold voltage, which will result in an exponential increase in I_{off} due to subthreshold leakage currents, thus also increasing P_{diss}. If the I_{on}/I_{off} ratio is fixed, V_{dd} can only be reduced if the steepness $\mathrm{d}I_d/\mathrm{d}V_{gs}$ of the device's I_d-V_{gs} characteristics can be increased.

There is a fundamental limitation to how steep the subthreshold region of a conventional MOSFET can be [146], which has been called the "Boltzmann-tyranny" [147]. Due to the thermal broadening of the distribution of electron energies in the source (approximately a Boltzmann distribution), the so-called subthreshold swing can be defined as [4]

$$S = \frac{\partial V_{gs}}{\partial (\log_{10} I_d)} = \frac{\partial V_{gs}}{\partial \psi_s} \frac{\partial \psi_s}{\partial (\log_{10} I_d)} = \ln(10) \frac{k_B T}{q} \left(1 + \frac{C_s}{C_{ins}}\right), \tag{2.32}$$

where ψ_s is the semiconductor surface potential, k_B is the Boltzmann constant, q is the elementary charge, C_s is the semiconductor capacitance and C_{ins} is the capacitance of the gate insulator. For any conventional MOSFET, where $C_{ins} > 0$, the subthreshold swing S must be larger than 60 mV decade^{-1} at room temperature ($T = 300$ K). However, if C_{ins} in Eq. 2.32 would be negative, S could be smaller than this fundamental limit, since then the so-called body factor $\partial V_{gs}/\partial \psi_s = 1 + C_s/C_{ins} < 1$. This means, using a gate insulator with NC, the surface potential ψ_s can be amplified with respect to V_{gs}. This *differential* voltage amplification ($\partial \psi_s/\partial V_{gs} > 1$) from passive components only is the main reason why ferroelectric NC is promising for applications in electronics. Besides ferroelectrics, it has been proposed that NC could be achieved by using an air gap [121, 122] or a piezoelectric material as the gate insulator [148]. Other proposed ways to overcome the Boltzmann tyranny without NC are e.g. band-to-band tunneling [149], impact ionization [150], band modulation by carrier injection [151] and thyristor-like structures [152]. Here, only the concept of a ferroelectric NCFET will be discussed.

Note that to avoid a hysteresis in the I_d-V_{gs} curve, the body factor must always be larger than zero (which yields a similar condition as Eq. 2.30). Nevertheless, a body factor below one might still be achieved when hysteresis is present due to the transient NC effect discussed previously. For device applications with the goal of a reduction of V_{dd} by using NC, hysteresis is undesirable. Indeed, ferroelectric field-effect transistors (FeFETs) have long been investigated for non-volatile data storage due to the hysteresis in their I_d-V_{gs} characteristics and non-destructuive memory read-out [153]. Therefore, it is sometimes not easy to draw a line between the FeFET and the NCFET concept, since both device structures are generally the same [154] and both can have $S < 60\,\mathrm{mV\,decade^{-1}}$ [155]. A further complication is the fact that many FeFETs also exhibit charge trapping, which might lead to an opposite hysteresis than the ferroelectric switching [156]. Under certain conditions, the overall hysteresis can almost vanish due to compensating ferroelectric switching and charge trapping [157]. Even charge trapping itself can lead to $S < 60\,\mathrm{mV\,decade^{-1}}$ [158]. However, a true NCFET should have no hysteresis even without charge trapping and for a broad range of operating conditions, e.g. different measurement speeds and voltage ranges.

The first report of a MOSFET with a ferroelectric gate insulator achieving $S < 60\,\mathrm{mV\,decade^{-1}}$ (and significant hysteresis) was reported in 2008 [159]. Afterwards, many reported devices applied an internal metal gate between the ferroelectric and a dielectric interface layer, thus enabling separate measurements of the device with and without ferroelectric [160]. However, the internal metal design seems to have some inherent drawbacks when it comes to NC stabilization, which will be investigated in Chapter 5. Many publications have since focused on optimizing the NCFET device design, often based on the single-domain Landau model, see e.g. [143, 161, 162, 163, 164, 165, 166, 167, 168, 169, 170, 171, 172, 173, 174, 175, 176]. Furthermore, many experimental investigations have shown that $S < 60\,\mathrm{mV\,decade^{-1}}$ is achievable in ferroelectric devices with varying degrees of hysteresis [177, 178, 179, 180, 181, 182, 183, 184, 185, 186, 187, 188, 189] and some even reporting negligible hysteresis under certain measurement conditions, see e.g. [190, 191, 192, 193, 194, 195, 196, 197, 198, 199, 200, 201]. Notably, for devices with smaller reported I_d-V_{gs} hysteresis, the improvement of S was also smaller [124, 202]. In fact, most devices described as "hysteresis-free" typically are very close to the $60\,\mathrm{mV\,decade^{-1}}$ limit or exhibit hysteresis when slightly changing the measurement conditions. At the time of writing, there has been no report of an experimental NCFET which shows S significantly below $60\,\mathrm{mV\,decade^{-1}}$ and no hysteresis for a wide range of measurement conditions at the same time. In this regard, a better theoretical understanding of the device design is needed [143]. On the other hand, a more fundamental understanding of NC in ferroelectric materials, and especially fluorite-type ferroelectrics, must be developed, which was one of the main goals of this thesis.

3 Experimental methods

In this chapter, the experimental procedures applied during this Ph.D. work are explained and discussed with respect to their applicability and potential limitations. Beginning with the description of the sample fabrication processes, the different applied thin film deposition techniques as well as annealing and patterning processes are described. Subsequently, the physical characterization techniques are briefly introduced, focusing on X-ray diffraction (XRD) and transmission electron microscopy (TEM) methods. Lastly, the different applied methods for electrical characterization of the samples are described.

3.1 Capacitor fabrication

For the experimental investigation and characterization of NC in ferroelectric materials, planar thin film capacitor structures were fabricated, containing insulating different ferroelectric and/or dielectric layers. The fabrication of such capacitors was achieved by applying different thin film deposition techniques for metal electrodes and insulating layers. Furthermore, to crystallize the used HfO_2 based thin films into their ferroelectric orthorhombic phase, thermal post-deposition annealing processes were applied. Finally, the samples were patterned into capacitor structures for electrical characterization. The general process flow is shown schematically in Fig. 3.1.

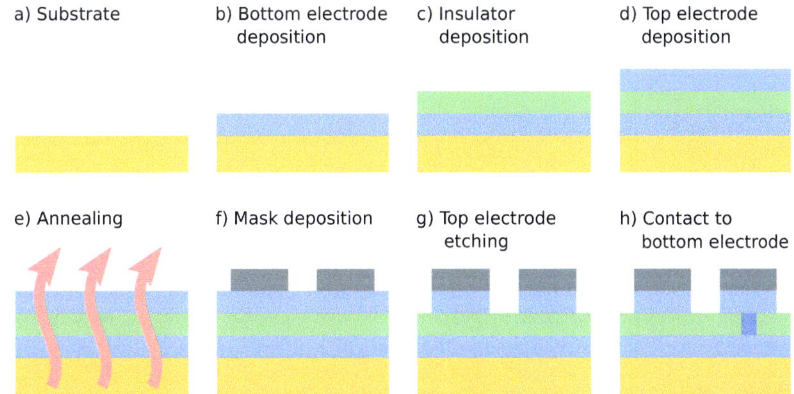

Figure 3.1: Schematic fabrication process for HfO_2 based ferroelectric capacitors used in this thesis. First, a metal-insulator-metal stack is deposited onto a substrate a)-d). Then, a thermal annealing treatment is applied as shown in e). In f) a metal is deposited as a hard mask and the top electrode is subsequently etched in g). Lastly, in h) electrical contact to the bottom electrode is achieved via dielectric breakdown of a capacitor.

As a substrate for the capacitors, standard commercial silicon wafers were used. First, conducting TiN bottom electrodes were deposited on the substrates. In a next step, insulating (ferroelectric and/or dielectric) layers were grown on top of the bottom electrodes as shown in Fig. 3.1c). Subsequently, TiN top electrodes were deposited, after which the samples were annealed at high temperature in a nitrogen atmosphere. To define the capacitor top electrodes, a metallic Ti/Pt hard mask was then deposited on top of the TiN, where the Ti served as an adhesion layer for the Pt. To insulate adjacent capacitors from each other, the TiN top electrode was then etched back, using the Pt as a hard mask. Lastly, the electrical contact to the bottom electrode was achieved by applying a high voltage between two capacitor top electrodes resulting in hard dielectric breakdown of the insulator. The top electrode of one of these broken capacitors was then used a contact to the bottom electrode.

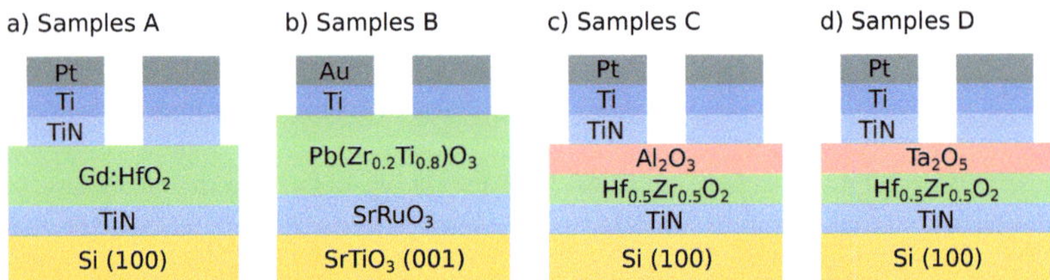

Figure 3.2: Schematic structures of the ferroelectric samples used in this thesis. Fabrication details can be found in Table 3.1.

Fig. 3.2 shows the schematic structure of the four main sample series investigated in this thesis. The first set of samples sketched in Fig. 3.2a) uses 3.4 cat% Gd:HfO$_2$ ferroelectric layers of 18 and 27 nm thickness between two TiN metal electrodes. The second sample B, on the other hand, consists of 100 nm ferroelectric Pb(Zr$_{0.2}$Ti$_{0.8}$)O$_3$ grown epitaxially on a SrRuO$_3$ buffered SrTiO$_3$ substrate. The third set of samples C consists of ferroelectric/dielectric heterostructures of Hf$_{0.5}$Zr$_{0.5}$O$_2$ (HZO) and Al$_2$O$_3$ with different thicknesses. Lastly, the samples D use a slightly different ferroelectric/dielectric heterostructure with Ta$_2$O$_5$ instead of Al$_2$O$_3$. The detailed processing conditions for all samples are summarized in Table 3.1.

Samples A were fabricated at imec by Christoph Adelmann and Mihaela Popovici until TiN top electrode deposition. The devices were annealed and structured at NaMLab by Tony Schenk. Samples B were fabricated by Claudy Serrao and Zhongyuan Lu at the University of California, Berkeley. Samples C were fabricated by Benjamin Max and myself at NaMLab. Samples D were fabricated by Franz Paul Gustav Fengler at NaMLab with help from Melanie Herzig and Terence Mittmann.

Table 3.1: Fabrication process conditions and layer thicknesses of the samples investigated in this thesis. RT: room temperature, ALD: atomic layer deposition, PLD: pulsed laser deposition, TEMAHf: Tetrakis(ethylmethylamino)hafnium, TEMAZr: Tetrakis(ethylmethylamino)zirconium, TMA: Trimethylaluminum.

	Samples A	Samples B	Samples C	Samples D
Substrate	Si (100)	$SrTiO_3$ (001)	Si (100)	Si (100)
Bottom electrode	12 nm TiN; reactive sputtering at RT	20 nm $SrRuO_3$; PLD at 630 °C	12 nm TiN; reactive sputtering at RT	12 nm TiN; reactive sputtering at RT
Ferroelectric	18 and 27 nm 3.4 cat% Gd:HfO_2; ALD at 300 °C: $HfCl_4$ & H_2O, Gd(^{i}PrCp)$_3$ & H_2O	100 nm Pb($Zr_{0.2}Ti_{0.8}$)O_3; PLD at 720 °C, 100 mTorr O_2 partial pressure	7.7 and 11.3 nm $Hf_{0.5}Zr_{0.5}O_2$; ALD at 260 °C: TEMAHf & H_2O, TEMAZr & H_2O	11.6 nm $Hf_{0.5}Zr_{0.5}O_2$; ALD at 260 °C: TEMAHf & H_2O, TEMAZr & H_2O
Dielectric (optional)			0.5 - 4 nm Al_2O_3; ALD at 260 °C TMA, H_2O	13.5 nm Ta_2O_5; reactive sputtering at RT
Top electrode	12 nm TiN; reactive sputtering at RT		12 nm TiN; reactive sputtering at RT	12 nm TiN; reactive sputtering at RT
Annealing	650 °C, 10 min in N_2 atmosphere	-5 °C min^{-1} cooling from 720 °C in 1 atm O_2	600 °C, 20 s in N_2 atmosphere	500 °C, 20 s in N_2 atmosphere
Contacts	10 nm Ti and 30 nm Pt; evaporation at RT	Ti/Au evaporation at RT	10 nm Ti and 30 nm Pt; evaporation at RT	10 nm Ti and 30 nm Pt; evaporation at RT
Patterning	Pt shadow mask; TiN etching in NH_4OH, H_2O_2, H_2O solution	lithography	Pt shadow mask; TiN etching in NH_4OH, H_2O_2, H_2O solution	Pt shadow mask; TiN etching in NH_4OH, H_2O_2, H_2O solution

3.1.1 Thin film deposition

For the fabrication of the capacitors, different thin film deposition methods were applied. While the metal electrodes were deposited by physical vapor deposition, most of the insulating layers were grown by atomic layer deposition (ALD). In the following, the basic principle of the different deposition techniques will be described.[203]

3.1.1.1 Physical vapor deposition

The term physical vapor deposition is used for a variety of vacuum deposition techniques where the material to be deposited is first transformed into a vapor phase and then condensates on a substrate. For the samples used in this thesis, the physical vapor deposition methods of sputtering, evaporation and pulsed laser deposition (PLD) were applied.[204]

Sputtering

In a sputter deposition tool, the sample, which should be coated, and a sputter target, which consists of the material that should be deposited, are facing each other in a vacuum chamber. Between the sample and the target, a large voltage is applied such that the target has a negative potential. An inert process gas, typically Argon (Ar), is let into the chamber such that a plasma of Ar^+ and e^- is formed between sample and target. Due to the potential difference, the Ar^+ ions are accelerated towards the target and ablate the material into the gas phase, which finally condensates on the sample surface. To enhance the density of Ar^+ ions in the plasma and therefore the sputter rate, permanent magnets are placed at the back of the target. Film deposition using such permanent magnets is called magnetron sputtering. In reactive sputtering, an additional gas is introduced into the vacuum chamber, which reacts with the sputtered material in the gas phase and is deposited on the sample. For example, N_2 can be used when sputtering a Ti target to deposit TiN, as was done for the electrodes of samples A, C and D. For the sputtering of insulating materials, radio frequency excitations can be used to prevent charge accumulation in the target. This was used when sputtering the dielectric Ta_2O_5 for samples D.

Evaporation

Deposition by evaporation is carried out under even lower pressure than sputtering, i.e. in the range of 10^{-6} Torr compared to 10^{-2} Torr for sputtering. The material which should be deposited is heated to very high temperatures, e.g. by an electron beam or a resistive heater, such that atoms are evaporating into the chamber and condensate on the sample. Disadvantages are poor step coverage and low growth rates. Nevertheless, electron beam evaporation was used here for the deposition of thin Ti and Pt layers on a flat substrate, where these disadvantages are negligible.

Pulsed laser deposition

Similar to evaporation, PLD utilizes a high power pulsed laser beam which is focused on a material target which should be deposited on a sample. The target material is vaporized into the vacuum chamber, which might be filled with a reactive process gas like O_2. The vaporized material, which exists in a plasma state is then deposited on the sample as a thin film. This method is typically used to epitaxially grow ferroelectric perovskite heterostructures with high quality interfaces on lattice matched substrates. PLD was used to grow epitaxial PZT on $SrRuO_3$ buffered $SrTiO_3$ (samples B).

3.1.1.2 Atomic layer deposition

ALD is a technique for the conformal deposition of ultra-thin films based on sequential, self-limiting chemical surface reactions [205, 206]. A typical ALD process cycle involves the following steps:

1. Precursor A chemically reacts with the substrate until a full coverage of the surface is achieved.

2. Excess species of precursor A and reaction byproducts are purged away from the substrate.

3. Precursor B chemically reacts with the surface ligands of precursor A until a full coverage with one (sub-)monolayer of the desired material is achieved.

4. Excess species of precursor B and reaction byproducts are purged away from the substrate.

Since the chemical surface reactions in step 1 and 3 are self-limiting, repeating steps 1 to 4 for multiple cycles leads to a controllable film thickness, which is given by the growth per ALD cycle times the number of ALD cycles. A typical growth per cycle for a self-limiting ALD process is in the range of 1 Å/cycle, which is in the order of magnitude of a single atomic monolayer. Therefore, with ALD it is possible to deposit extremely thin films with precisely defined composition and thickness even on complex 3D structures. The main disadvantages of ALD are the low deposition rate and expensive equipment and chemicals needed.

In the samples C and D, which were fabricated at NaMLab, an Oxford Instruments OpAL ALD tool was used to grow thin ferroelectric HZO and dielectric Al_2O_3 layers. For the former, alternating ALD cycles of TEMAHf/water and TEMAZr/water were used to achieve an equal mixture of HfO_2 and ZrO_2 at a deposition temperature of 260 °C. Al_2O_3 layers were deposited by using TMA and water as precursors also at 260 °C substrate temperature. On the other hand, $Gd:HfO_2$ films were grown by thermal ALD at imec using $Gd(^iPrCp)_3$/water and $HfCl_4$/water as precursors at 300 °C. In this case, the ALD cycle ratio between HfO_2 and Gd_2O_3 was 27:1, resulting in 3.4 % of Gd content in the films.

3.1.2 Thermal annealing

To obtain ferroelectric HfO_2 based films, the material must be crystallized into the non-centrosymmetric orthorhombic $Pca2_1$ phase [207]. Since the ALD deposited HZO and $Gd:HfO_2$ films were amorphous as deposited and also stayed amorphous after deposition of the TiN top electrode, another thermal annealing process was necessary to crystallize the films into the orthorhombic phase. Previous studies have shown, that the crystallization into the orthorhombic phase is facilitated by first depositing the top electrode, a film thickness range of around 10 nm and an appropriate amount of doping [8, 91, 142]. For samples A using $Gd:HfO_2$, an annealing temperature of 650 °C was used for 10 minutes in a nitrogen atmosphere. Since HZO typically has a lower crystallization temperature [82], 500 °C and 600 °C for 20 s were sufficient for samples D and C, respectively. The epitaxial perovskite ferroelectric samples B did not require additional annealing as they were already crystalline during the high temperature PLD process.

3.1.3 Capacitor patterning

For samples A, C and D, the patterning of the capacitors was achieved by evaporating the Ti/Pt contacts through a shadow mask. Subsequently, the TiN top electrode was etched back using a standard-clean (SC-1) solution consisting of NH_4OH, H_2O_2 and H_2O in a ratio of 1:2:50 at 50 °C. This etching solution is generally selective to oxide layers, which is why the etching will stop once the TiN is removed and the underlying oxide film is exposed [208]. For samples B, standard lithographic patterning techniques to structure the Ti/Au contacts were used instead.

3.2 Physical characterization

The main physical characterization methods applied in this work were grazing-incidence X-ray diffraction (GIXRD) and TEM. These techniques are useful to determine the crystal structure and morphology of thin polycrystalline layers and were therefore applied to structurally characterize the thin ferroelectric HfO_2 based layers.

3.2.1 Grazing-incidence X-ray diffraction

GIXRD is a crystallographic technique which is very sensitive to the crystal structure on the surface of a sample [209]. Therefore, it is often used to characterize polycrystalline thin films, such as ferroelectric HfO_2 based ones [97, 142]. In an experimental GIXRD setup, a beam of X-rays is shone on the sample with a constant and very small incident angle $\sim 0.45°$ with respect to the sample surface. The incident X-rays are then diffracted by the differently oriented grains inside the polycrystalline film under various different angles. To capture these various diffraction beams, the X-ray detector is then scanned along the 2θ angle. The resulting diffraction pattern can then be compared to theoretical reference patterns for the different expected crystal phases in the material. Based on the crystal symmetry, some combinations of diffraction peaks at certain angles of 2θ can be used to distinguish different phases in the polycrystalline layer. However, in the case of HfO_2 based ferroelectrics, especially the ferroelectric orthorhombic $Pca2_1$ and the non-ferroelectric tetragonal $P4_2/nmc$ phase have very similar lattice constants complicating an unambiguous distinction of their diffraction patterns [97, 142]. In this work, GIXRD was carried out on a Bruker D8 Discover using Cu-Kα radiation with a wavelength $\lambda = 0.154$ nm.

3.2.2 Transmission electron microscopy

TEM is a technique to image the structure of a thin lamella of material with atomic resolution through the interaction of electrons with very high energy ($\sim$100-300 keV), i.e. small wavelength. Since the wavelength of these electrons is orders of magnitude smaller than visible light, much more detailed features can be resolved using TEM. However, to achieve highest resolution, precise sample preparation is necessary. Typically, thin lamellas ($\sim$100 nm) are fabricated using a focused ion beam. Then the high energy electrons are focused on this lamella and measured on the other side after being transmitted through the sample.[210]

3.2.2.1 Selected area electron diffraction

To determine the crystal structure of the sample, selected area electron diffraction (SAED) is a method were only a small area of the lamella is hit by the electron beam by using an aperture to block the beam for the rest of the sample. From the incident high energy electrons, some fraction will be diffracted. This will lead to a specific electron diffraction pattern based on the local crystal structure [210]. In contrast to GIXRD, this technique as a much higher lateral resolution due to the small electron wavelength.

3.2.2.2 Electron energy loss spectroscopy

Another technique which can be used in conjunction with TEM to obtain more information about a sample is electron energy loss spectroscopy (EELS). When a beam of electrons with a narrow energy distribution is used, it is possible to measure the energy loss of the electrons due to inelastic scattering processes in the sample. Since some of these inelastic scattering processes like the ionization of inner electron shells are element-specific, EELS can be used to obtain information on the local elemental composition of the sample.[210]

3.3 Electrical characterization

One of the main focuses of this thesis was the electrical characterization of ferroelectric(/dielectric) capacitors and related NC phenomena. In the following, dynamic hysteresis measurements, small-signal capacitance-voltage measurements and pulsed charged voltage measurements are described.

3 Experimental methods

3.3.1 Dynamic hysteresis measurements

A standard electrical characterization technique for ferroelectric capacitors is the dynamic measurement of the polarization-electric field hysteresis, often called dynamic hysteresis measurement. This method is based on the application of several triangular voltage signals, from which many important ferroelectric parameters like remanent polarization or coercive field can be calculated. The general applied voltage waveform is shown in Fig. 3.3, where f is the measurement frequency and V_{max} is the maximum voltage amplitude. Typical measurement frequencies are in the range of 100 Hz to 10 kHz.

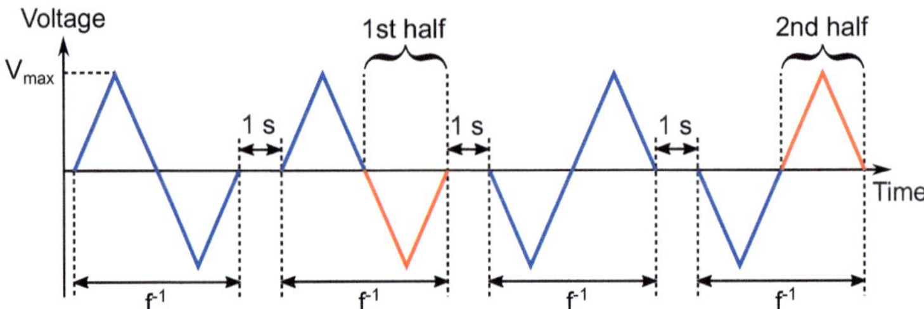

Figure 3.3: General voltage waveform used for dynamic hysteresis measurements of ferroelectric capacitors. V_{max} is the maximum voltage amplitude and f is the measurement frequency. Red segments indicate the times during which the closed hysteresis loop is acquired.

Between the four triangular signals, a delay of one second is inserted, which enables the calculation of the relaxed remanent polarization, which is an indicator for short term retention. The actual closed loop polarization-electric field hysteresis is obtained by combining the second half of the second triangular signal and the second half of the fourth signal, as indicated by the red color in Fig. 3.3. However, since the ferroelectric polarization response $P(t)$ cannot be directly measured, it has to be calculated from the measured current $I(t)$ as

$$P(t) \approx \frac{Q(t)}{A} = \frac{1}{A} \int I(t)\mathrm{d}t + P_{offset}, \tag{3.1}$$

where the integration constant P_{offset} is chosen such that the maximum and minimum polarization P_{max} and P_{min} obey $P_{max} = -P_{min}$, thus centering the P-E_f loop along the P-axis. Note that the measured polarization in this case includes both the spontaneous polarization P_S and the background polarization P_b. The electric field in the ferroelectric (for an MFM capacitor) is given by $E_f(t) = V(t)/t_f$. Dynamic hysteresis measurements were carried out on a Keithley 4200 Semiconductor Characterization System equipped with 4225 Pulse/Measure Units and remote amplifier.

3.3.2 Small-signal capacitance-voltage measurements

In a small-signal capacitance-voltage (*C-V*) measurement, a small AC voltage signal $v_{AC}(t) = V_{AC}\sin(\omega t)$ is superimposed on a DC voltage sweep, as indicated in Fig. 3.4. Here, V_{AC} is the small-signal voltage amplitude, $\omega = 2\pi f$ is the angular frequency and f is the frequency. For each DC voltage step V_{step}, the small-signal current response is measured over the time T_{step}. The longer T_{step}, the better the signal-to-noise ratio, since the measured capacitance is obtained from the integrated current response over time. Since V_{AC} is typically in the range of 50 mV, the current response can be very small, making a long T_{step} necessary. Typically, T_{step} is in the range of tens of milliseconds. Therefore, such small-signal *C-V* measurements are rather slow, especially when measuring capacitances below 1 nF as done in this work.

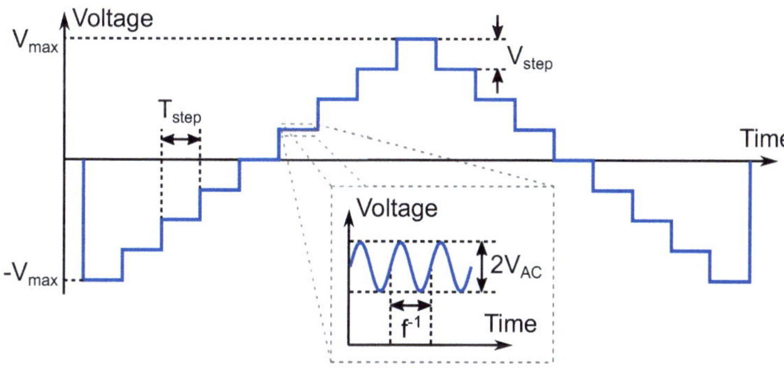

Figure 3.4: Schematic applied voltage signal for small-signal capacitance-voltage measurements. V_{max}, V_{step}, T_{step}, f and V_{AC} are the maximum voltage amplitude, voltage step size, time step size, small-signal frequency and small-signal voltage amplitude, respectively.

While for an ideal capacitor, the phase shift of the current response compared to the voltage signal would be $\pi/2$, in real materials it is less than $\pi/2$, due to loss effects, e.g. leakage currents. To consider these non-idealities, the sample can be modeled as an equivalent resistance R_{epr} in parallel with an ideal capacitance C. Then, the loss factor can be defined as $\tan(\delta) = 1/(\omega R_{epr}C)$, where δ is the loss angle, which is zero for an ideal capacitor.

In the case of ferroelectric capacitors, the long measurement time inevitably leads to hysteresis in the *C-V* graph due to ferroelectric switching if the DC voltage range (given by V_{max}) is large enough. In this case, ferroelectrics exhibit 'butterfly'-shaped *C-V* characteristics. For many applications, small-signal *C-V* measurements are sufficient to characterize the electrical behavior of the device. However, in other cases, the measurement is too slow, leading to other undesirable effects like leakage currents and charge injection being superimposed on the measured data [211]. Furthermore, faster transient charging and discharging effects of non-linear capacitors like ferroelectrics cannot be captured by small-signal *C-V* methods. Therefore, fast pulsed charge-voltage methods have been developed, which are described in the following.

3.3.3 Pulsed charge-voltage measurements

To characterize the transient electrical response of a device, short voltage pulses can be applied using a pulse generator, while measuring the transient current response through a small series resistance R [57, 127]. To obtain a sufficiently high time-resolution for the current measurement, the use of an oscilloscope is necessary, which measures the voltage drop V_R across the resistance R, see Fig. 3.5a). The device under test (DUT), here a capacitor, has the charge Q on its plates. By measuring the current $I = V_R/R$, the change in charge on the DUT can be calculated by integrating the current with respect to time, see Eq. 3.1.

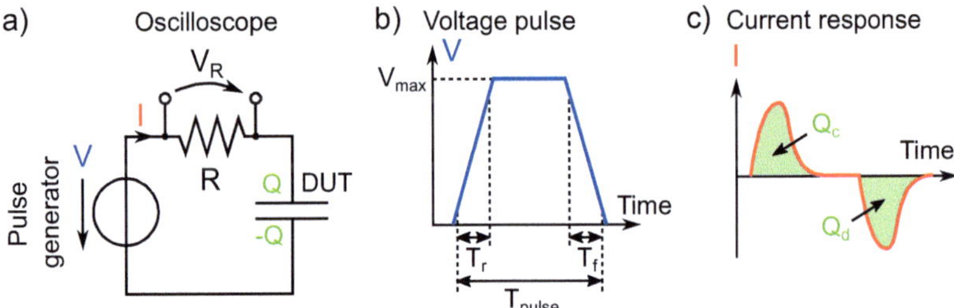

Figure 3.5: a) Simplified circuit diagram for pulsed charge-voltage measurements (Q-V). I is the current flowing through the resistor R, resulting in a voltage drop V_R. DUT is the device under test. b) Definition of the applied voltage pulse with pulse amplitude V_{max}, rise time T_r, fall time T_f and pulse length T_{pulse}. c) Transient current response, where the accumulated charge flowing during charging (Q_c) and discharging (Q_d) are indicated by the green areas.

Fig. 3.5b) shows an exemplary applied voltage pulse V, which is defined by the amplitude V_{max}, the pulse width T_{pulse} and the rise and fall times T_r and T_f, respectively. Both T_r and T_f are typically defined from 10 % to 90 % of the voltage amplitude. The capacitive current response to such a voltage pulse is shown in Fig. 3.5c). Beginning with the rising voltage, a positive current is flowing which drops to zero once the capacitor is fully charged at the voltage V_{max}. Once the applied voltage is reduced, a negative current is flowing, thus discharging the capacitor.

The area below the current peaks corresponds to the changes in charge during charging (Q_c) and discharging (Q_d). For an ideal capacitor, $Q_c = Q_d$. However, in real capacitors, leakage currents and charge trapping can lead to $Q_c \neq Q_d$. This pulsed measurement approach can also be used to measure the capacitance as a function of voltage: By applying consecutive voltage pulses with changing V_{max}, and measuring Q_d for each pulse. The differential capacitance can then be calculated as

$$C = \frac{\Delta Q_d}{\Delta V_{max}}. \tag{3.2}$$

The reason why Q_d should be used is because Q_c can be larger than Q_d when leakage currents are flowing. These leakage currents are largely negligible during discharging, because the applied voltage is much lower compared to the charging process. This method was originally proposed by Kim et al.[211]

Furthermore, the transient voltage across the DUT can be measured or calculated as $V_{DUT} = V - IR$. In this way, the capacitance can even be determined in the time domain, by analyzing dQ/dt and dV_{DUT}/dt as a function of time, which was suggested by Khan et al.[127]

4 Transient negative capacitance in metal-ferroelectric-metal capacitors

In order to directly measure NC in a ferroelectric, it is necessary to simultaneously measure the changes in the voltage across the ferroelectric V_F as well as the charge Q_F on its surface, since the ferroelectric capacitance is defined as $C_F = dQ_F/dV_F$. This is only possible when integrating the ferroelectric into an MFM capacitor, since otherwise the voltage drop across the ferroelectric cannot be measured directly, e.g. with an oscilloscope. Furthermore, to obtain the change in Q_F, a second circuit element in series with the MFM capacitor is needed, which could be either a passive device like a resistor, a capacitor (Fig. 4.1a,b)) or an active device like a MOSFET (Fig. 4.1c)) [124]. However, this direct measurement approach has the limitation that hysteresis effects are practically unavoidable when the ferroelectric is switching, due to the small screening length of the metal electrodes.

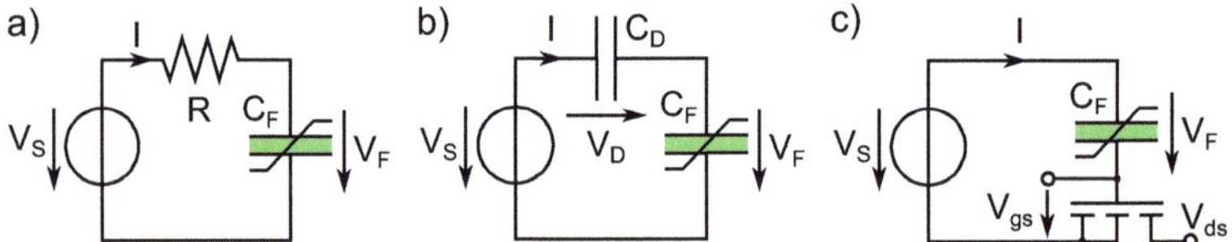

Figure 4.1: The three main circuit configurations for direct measurements of NC. In principle, the voltage source V_S can be replaced by a current source. All configurations consist of a ferroelectric capacitor C_F which is in series with a) a resistor R, b) a dielectric capacitor C_D or c) the gate electrode of a MOSFET. I, V_F, V_D, V_{gs} and V_{ds} are the current, the voltage across the ferroelectric, the voltage across the dielectric, the gate-source voltage and the drain-source voltage, respectively.

As first proposed and demonstrated by Khan et al. [127], the simplest way to measure changes in Q_F and V_F at the same time is to connect a resistance R in series to the MFM capacitor while applying a voltage V_S as shown in Fig. 4.1a). By measuring the voltage across R, the current I can be calculated by Ohm's law. Finally, the change in Q_F is obtained by integrating I over time. In contrast, when connecting a dielectric capacitor C_D in series as shown in Fig. 4.1b), Q_F is directly proportional to the voltage V_D across C_D. Another way to think about this is in terms of a differential voltage amplification, which can be defined as $A_V = dV_D/dV_S$. For positive capacitors in series, $A_V < 1$ is always true. However, if $A_V > 1$ then the ferroelectric must have a NC. Similarly, when connecting an MFM capacitor to the gate of a MOSFET as shown in Fig. 4.1c), then NC will result in $A_V = dV_{gs}/dV_S > 1$, which can reduce the subthreshold swing below the Boltzmann limit of $2.3k_BT/q$.[4]

The outline of this chapter is as follows: First, the direct measurement of transient NC in a polycrystalline layer of ferroelectric doped HfO_2 is described. Secondly, the ferroelectric domain switching dynamics during transient NC measurements are investigated in epitaxial PZT capacitors as a ferroelectric model system. Lastly, the direct observation of differential voltage amplification due to transient NC in a circuit with a ferroelectric and dielectric capacitor in series is demonstrated.

4.1 Transient negative capacitance in ferroelectric hafnium oxide

While the first direct measurement of transient NC reported by Khan et al. [127] was an important step towards proving the concept of NC in general, it was measured using an epitaxial ferroelectric PZT film, which is extremely difficult to integrate into highly-scaled MOSFET devices. Therefore, a further demonstration that CMOS compatible and polycrystalline ferroelectric HfO_2 based thin films also could exhibit NC was needed. To achieve this, a similar measurement setup as shown in Fig. 4.1a) was used in combination with high-quality ferroelectric $Gd:HfO_2$ capacitors [212, 213]. This section is based on the results published in Ref. [137].

4.1.1 Sample structure

Fig. 4.2 shows the structural properties of the $TiN/Gd:HfO_2/TiN$ MFM capacitors used in these experiments. While the TEM cross-sections in Fig. 4.2a,b) prove the polycrystalline nature of the ferroelectric and electrode films, the GIXRD patterns in Fig. 4.2c) reveal the phase composition for $Gd:HfO_2$ layers of 18 nm and 27 nm thickness.

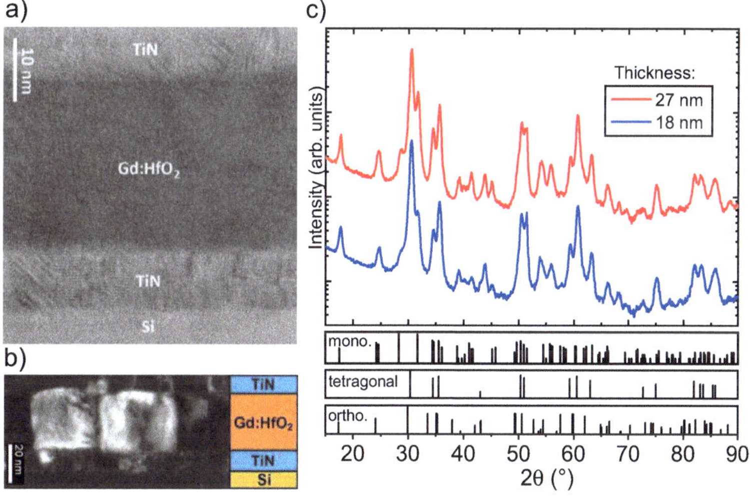

Figure 4.2: a) TEM cross-section of a 27 nm $Gd:HfO_2$ capacitor. b) Dark-field TEM cross-section shows columnar HfO_2 grain structure. c) GIXRD data of 18 and 27 nm thin $Gd:HfO_2$ films with reference patterns of the monoclinic, tetragonal, and orthorhombic phases. Copyright 2016 Wiley. Adapted with permission from Ref. [137].

The dark-field TEM cross-section in Fig. 4.2b) shows that the grain size of the Gd:HfO$_2$ is comparable to its film thickness, which is in agreement to other reports on HfO$_2$ based ferroelectrics [92, 142, 214, 215]. Furthermore, grain boundaries are generally observed perpendicular to the film plane, which results in a columnar grain structure [97]. The GIXRD patterns indicate a mixture of monoclinic ($P2_1/c$), tetragonal ($P4_2/nmc$) and orthorhombic ($Pca2_1$) phases in the Gd:HfO$_2$ films. While a distinction between the ferroelectric orthorhombic and (antiferroelectric) tetragonal phase from GIXRD is generally challenging, a reduction of the monoclinic phase fraction for the 18 nm film compared to the 27 nm film is clearly visible. This trend can be explained by the reduced grain sizes in thinner films, which due to the surface energy effect favors the orthorhombic and tetragonal phases compared to the monoclinic one [100].

4.1.2 Electrical characterization

The GIXRD results are consistent with the P-E_f hysteresis loops measured at 10 kHz for the different Gd:HfO$_2$ thicknesses shown in Figure 4.3. For the thicker 27 nm film, the remanent polarization P_r is lower compared to the 18 nm sample (13 µC cm^{-2} vs. 16 µC cm^{-2}), which is reasonable due to the lower monoclinic and higher orthorhombic phase fraction in the latter. On the other hand, the coercive field E_c for the 27 nm film is only slightly lower (1.6 MV cm^{-1}) than for the 18 nm film (1.7 MV cm^{-1}). Overall, both P-E_f loops have rather steep flanks (dP/dE_f) around E_c compared to most reported ferroelectric HfO$_2$ capacitors, indicating abrupt switching behavior, which is helpful to observe transient NC [127]. In the following, the transient NC measurement methodology and results will be presented.

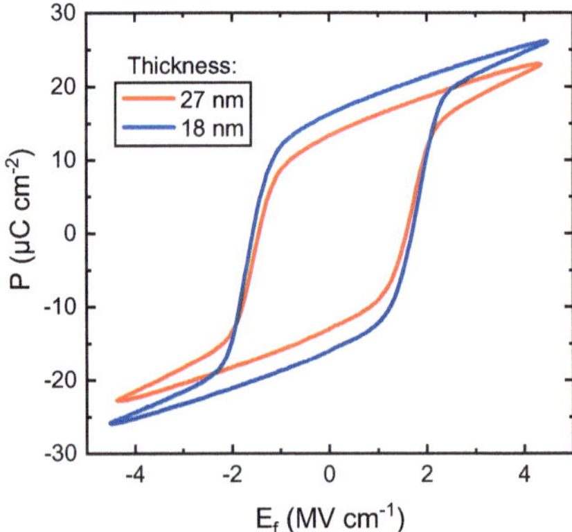

Figure 4.3: Polarization P versus electric field E_f hysteresis of ferroelectric Gd:HfO$_2$ capacitors with different thickness. Copyright 2016 Wiley. Adapted with permission from Ref. [137].

As schematically shown in Fig. 4.1a), the simplest way to observe transient NC is to connect an external resistor in series to the MFM capacitor and apply a voltage V_S. More specifically, Fig.

4.4 depicts the experimental setup used to directly observe NC in the Gd:HfO$_2$ capacitors. A Keithley 4200-SCS semiconductor characterization system equipped with Pulse/Measure Units and Remote Amplifier/Switches (4225-RPMs) was used to apply short ($\sim$µs) voltage pulses to the series connection of R and the MFM capacitor. Both the applied voltage V_S and the voltage across the ferroelectric V_F was measured via an Agilent DSO5054A oscilloscope. As indicated in Fig. 4.4 the contact to the TiN bottom electrode was achieved by connecting the top electrode of a second shorted MFM capacitor. All connecting cables were kept as short as possible to minimize parasitic capacitances.

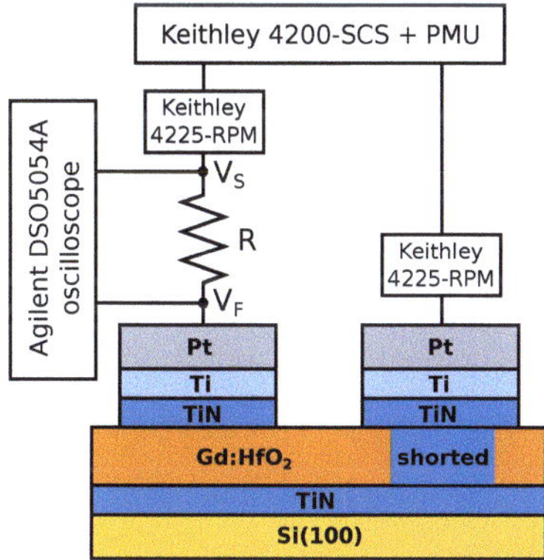

Figure 4.4: Schematic measurement setup and sample structure for measuring NC transients. V_S denotes the applied voltage whereas V_F represents the voltage across the ferroelectric. Copyright 2016 Wiley. Adapted with permission from Ref. [137].

As discussed by Khan et al. [127], transient NC can only be observed when the ferroelectric switches from one polarization state to the opposite one. Therefore, the Gd:HfO$_2$ was initially set into the negatively saturated polarization state by applying a large negative voltage V_S, before reversing the polarity and thus switching the ferroelectric into the positive polarization state. After the charging of the MFM capacitor was completed (when $V_S = V_F$), the applied voltage was again reversed to switch the ferroelectric back into the negative polarization state. Fig. 4.5 shows the transient voltage waveforms V_F and V_S for a 27 nm Gd:HfO$_2$ capacitor with an electrode area of $\sim$30.000 µm^2 and $R = 680\,\Omega$.

As can be seen in the top right inset in Fig. 4.5, there are three distinct regions in the charging transients of the switching ferroelectric capacitor. First, there is a fast dielectric charging region until $\sim$0.5 µs in which V_F increases according to the regular charging behavior of a positive linear capacitor. Then, from 0.5 µs to 1.5 µs V_F decreases due to the polarization switching, which is exactly opposite to what is expected for a positive capacitor. Finally, after a significant fraction of the polarization switching is completed, V_F is increasing again corresponding to a positive capacitance state. The same analogous behavior can be observed in the reverse switching process when applying the negative voltage pulse.

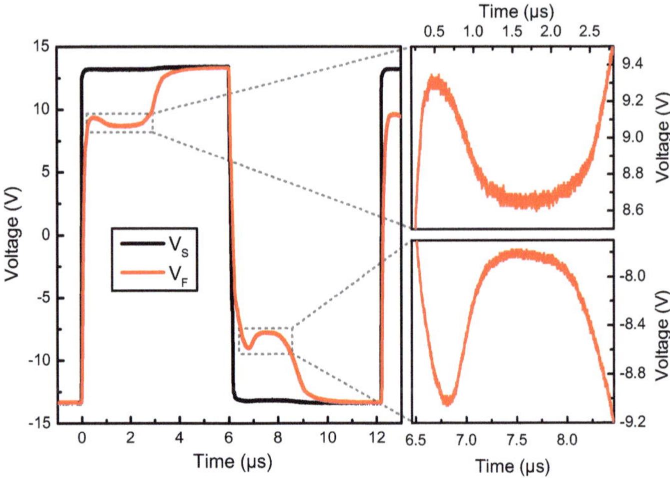

Figure 4.5: Voltage transients of a 27 nm Gd:HfO$_2$ capacitor with a series resistor R = 680 Ω. V_S is the applied voltage and V_F is the voltage across the ferroelectric. Insets on the top and bottom right hand side show magnifications of the NC regions. Copyright 2016 Wiley. Adapted with permission from Ref. [137].

To better understand why this transient change in V_F can be interpreted as a direct measurement of NC, it is instructive to plot the current flowing through the resistor I_R as well as the charge on the capacitor Q_F as a function of time. In general the capacitance of the MFM capacitor is defined as

$$C_F = \frac{dQ_F}{dV_F},\tag{4.1}$$

where C_F is a function of Q_F and therefore of V_F. Thus, if dQ_F/dt and dV_F/dt have opposite signs in a certain time frame, C_F must be negative according to Eq. 4.1 [127]. Now the current flowing through R can be calculated as

$$I_R = \frac{V_S - V_F}{R}.\tag{4.2}$$

To calculate Q_F the reasonable assumption can be made that I_R is much larger than the leakage currents of the MFM capacitor. However, the parasitic capacitance of the measurement setup C_{setup} has to be considered, which can be thought of as in parallel to C_F [127]. Therefore, the change of the total charge which is given by I_R, is actually the sum of the changes of Q_F and the charge on C_{setup}, i.e.

$$I_R = \frac{dQ}{dt} = \frac{dQ_F}{dt} + \frac{dV_F}{dt}C_{setup}.\tag{4.3}$$

By inserting Eq. 4.2 into Eq. 4.3 and integrating with respect to time, Q_F can be calculated as

$$Q_F(t) = Q_F(t=0) + \frac{1}{R} \int_0^t \left[V_S(t') - V_F(t') \right] \mathrm{d}t' - V_F(t) C_{setup}. \qquad (4.4)$$

Fig. 4.6 shows the transient behavior of V_S, V_F, I_R and Q_F for the same measurement as in Fig. 4.5. The dashed gray lines indicate where the sign of C_F changes. Between line A and B, it can be seen that V_F decreases while Q_F increases, which means that $C_F < 0$. Analogously, between line C and D, V_F increases while Q_F decreases, which also results in $C_F < 0$.

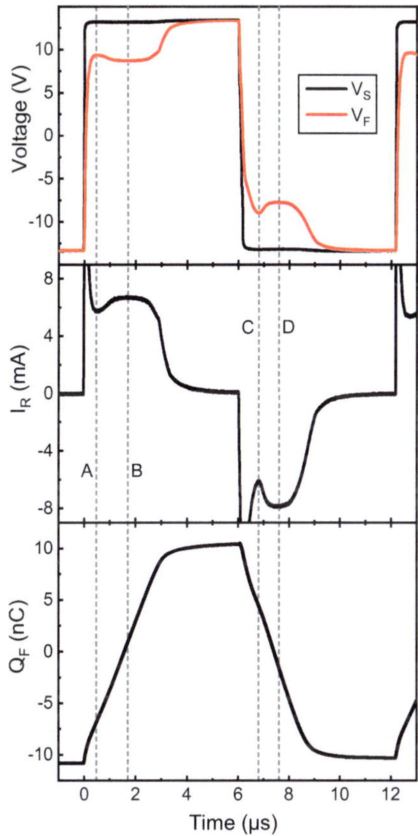

Figure 4.6: NC transients of a 27 nm Gd:HfO$_2$ capacitor with a series resistor R = 680 Ω. V_S is the applied voltage, V_F is the voltage across the ferroelectric, I_R is the current flowing through R and Q_F is the charge of the ferroelectric. Between the lines A and B as well as C and D, the capacitance is negative. Copyright 2016 Wiley. Adapted with permission from Ref. [137].

At all other times, $C_F > 0$, since V_F and Q_F change in the in the same direction. This shows that in this direct measurement, the NC state is only observed for a short time during the ferroelectric polarization switching and is associated with a significant hysteresis. The reason for this behavior is the fast screening of bound polarization charges by the free electrons in the metal electrodes [138]. This entails a destabilization of the NC state for longer timescales, e.g. in the samples described in this section after several microseconds.

However, when the bound polarization charge in the ferroelectric temporarily changes faster than the free screening charge on the electrodes, NC can be observed in such a transient measurement [138]. This is possible since the series resistance R slows down the supply of free charge to the electrodes and the change of the ferroelectric polarization during switching can be faster than the screening. This sets some important limitations for the usage of transient NC in large MFM capacitors, since 1) hysteresis cannot be avoided and 2) the timescale of the NC effect is ultimately limited by the switching speed of the ferroelectric. This fundamental limitation of the MFM structure for NC will be further elucidated in Section 4.2.

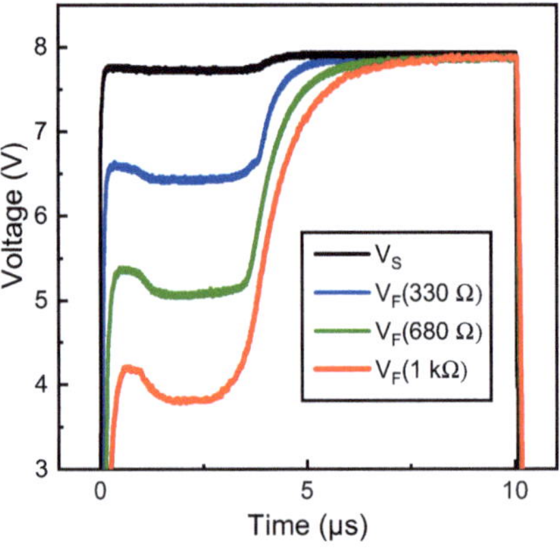

Figure 4.7: Voltage transients of a 18 nm Gd:HfO$_2$ capacitor measured using different series resistances R of 330 Ω (blue), 680 Ω (green), and 1 kΩ (red). With increasing series resistance the onset of the NC transient is observed at lower V_F. Copyright 2016 Wiley. Adapted with permission from Ref. [137].

Similar NC transients were observed for the 18 nm Gd:HfO$_2$ sample as shown in Fig. 4.7 for three different external resistors from 330 Ω to 1 kΩ. By increasing R, the voltage V_F at which the ferroelectric begins to switch is decreased significantly. At the same time, the decrease of V_F in the NC region is larger for higher resistance values. However, the overall duration of the NC state is similar for all three resistors. A qualitative explanation for this behavior will be presented in Section 4.2 in comparison to epitaxial PZT capacitors with distinctly different switching dynamics.

4.1.3 Multi-grain Landau-Khalatnikov simulations

To obtain further insights into the origin of the transient NC behavior, a physical model based on Landau-Khalatnikov theory consistent with the polycrystalline morphology of the Gd:HfO$_2$ films was developed [216]. This multi-grain Landau-Khalatnikov (MGLK) model is based on a few simplifying assumptions: The ferroelectric consists of N non-interacting columnar grains with an out-of-plane spontaneous polarization which is homogeneous inside each grain (single-domain approximation). Furthermore, only one-dimensional electrostatics in the out-of-plane direction of the film are considered. For simplicity, strain effects are not modeled and the background permittivity is set to zero. Under these constraints, according to LGD theory, the Gibb's free energy density G of each grain $j = 1, 2, ..., N$ can be written as

$$G_j = \alpha_j P_j^2 + \beta_j P_j^4 + \gamma_j P_j^6 - \left(E_f - E_{bias,j}\right) P_j, \qquad (4.5)$$

where α_j, β_j, γ_j are the LGD coefficients, P_j is the total polarization and $E_{bias,j}$ is the local internal bias field of the grain with index j. The total Gibb's free energy of each grain can then be defined as

$$U_j = G_j A_j t_f, \qquad (4.6)$$

where A_j is the effective capacitor area of the grain and t_f is the thickness of the ferroelectric. Accordingly, the ferroelectric charge on the capacitor for each grain is to a good approximation given by

$$Q_{Fj} = P_j A_j. \qquad (4.7)$$

Similarly, by multiplying the electric fields by the film thickness, V_F and the internal bias voltage V_{bj} are obtained from

$$V_F - V_{bj} = \left(E_f - E_{bias,j}\right) t_f. \qquad (4.8)$$

To simplify the following equations expressed in terms of charges and voltages, it is useful to define a new set of LGD coefficients as

$$a_j = \frac{2\alpha_j t_f}{A_j}, \qquad (4.9)$$

$$b_j = \frac{4\beta_j t_f}{A_j^3},\tag{4.10}$$

$$c_j = \frac{6\gamma_j t_f}{A_j^5}.\tag{4.11}$$

Now, the total Gibb's free energy of each grain j can be written as

$$U_j = \frac{a_j}{2}Q_{Fj}^2 + \frac{b_j}{4}Q_{Fj}^4 + \frac{c_j}{6}Q_{Fj}^6 - \left(V_F - V_{bj}\right)Q_{Fj}.\tag{4.12}$$

The Landau-Khalatnikov equation then yields the following:

$$R_{ij}\frac{\partial Q_{Fj}}{\partial t} = -\frac{\partial U_j}{\partial Q_{Fj}} = -\left(a_j Q_{Fj} + b_j Q_{Fj}^3 + c_j Q_{Fj}^5 - \left(V_F - V_{bj}\right)\right),\tag{4.13}$$

where R_{ij} is a newly introduced parameter of unit $[\Omega]$, which can be understood as an internal series resistance of each grain which is related to the resistivity parameter ρ_j from Eq. 2.21 by $R_{ij} = \rho_j A_j / t_f$. Furthermore, one can define

$$V_{ij} = a_j Q_{Fj} + b_j Q_{Fj}^3 + c_j Q_{Fj}^5 + V_{bj}\tag{4.14}$$

as the internal voltage of the jth grain. Therefore, one can express the ferroelectric voltage as

$$V_F = R_{ij}\frac{\partial Q_{Fj}}{\partial t} + V_{ij} = R_{ij}I_{Fj} + V_{ij},\tag{4.15}$$

where I_{Fj} is the ferroelectric current contribution of grain j. The equivalent circuit of the complete MGLK model can be seen in Fig. 4.8. From Kirchhoff's voltage law, the externally applied voltage V_S can be expressed as

$$V_S = I_R R + V_F,\tag{4.16}$$

with the current through the resistor as

$$I_R = \sum_{j=1}^{N} I_{Fj} + I_L + I_{Cp},\tag{4.17}$$

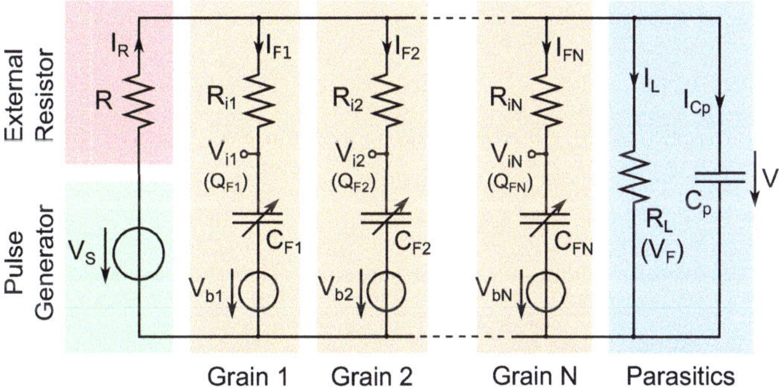

Figure 4.8: Equivalent electric circuit representing the multigrain Landau–Khalatnikov model for the ferroelectric including independent variables for each of the N grains: R_i is the internal resistance, V_i is the internal voltage, Q_F is the charge, C_F is the capacitance, I_F is the current, and V_b is the internal bias voltage of a grain. C_p and R_L are the parasitic capacitance and resistance, respectively and I_{Cp} and I_L are the currents flowing through them. The current flowing through R is I_R. Copyright 2016 Wiley. Adapted with permission from Ref. [137].

where I_{Cp} is the current flowing through the parasitic capacitor and I_L is the leakage current. The charge on the parasitic capacitor is given by

$$Q_{Cp} = V_F C_p. \tag{4.18}$$

It should be noted that C_p has two different contributions: C_{setup}, which is related to external influences like cable capacitances etc. and C_{sample}, which originates from the ferroelectric capacitor itself. Generally, non-ferroelectric grains in the Gd:HfO$_2$ as well as the dielectric background permittivity of the ferroelectric grains will contribute to C_{sample} and thus C_p. For the leakage current I_L, a generic exponential model is used according to

$$I_L = I_{L0}\text{sign}(V_F)\exp\left(\frac{|V_F|}{V_{L0}}\right), \tag{4.19}$$

where I_{L0} and V_{L0} are fitting parameters. Furthermore, the ferroelectric and parasitic charges are calculated by integrating the respective currents as

$$Q_{Fj}(t) = Q_{Fj}(t=0) + \int_0^t I_{Fj}(t')\mathrm{d}t', \tag{4.20}$$

$$Q_{Cp}(t) = Q_{Cp}(t=0) + \int_0^t I_{Cp}(t')\mathrm{d}t'. \tag{4.21}$$

As initial conditions it was assumed that no current is flowing which leads to

$$Q_{Cp}(t=0) = C_p V_S(t=0), \tag{4.22}$$

$$V_{ij}(t=0) = V_F(t=0) = V_S(t=0). \tag{4.23}$$

Lastly, the initial ferroelectric charge is set either as the positive or negative remanent polarization state as

$$Q_{Fj}(t=0) = Q_{F0j}. \tag{4.24}$$

Each ferroelectric grain is defined by a set of parameters (α, β, γ, ρ, E_{bias}) which determine its charge-voltage characteristics and switching speed. To emulate the physical variations of grain sizes and orientations, dopant fluctuations, defects and strains which have been observed in experiments, Gaussian distributions for all grain parameters are used for simplicity. Note that this MGLK model assumes that each individual grain switches homogeneously, which would make this transient NC effect in HfO_2 based ferroelectrics an intrinsic effect. It is still not clear if this is reasonable based on the experimental switching kinetics of individual grains, but recent studies indicate nucleation-limited switching kinetics [217]. To calibrate the MGLK model to experimental data, R is set to zero and standard current-voltage characteristics are simulated for a triangular V_S waveform with a frequency of 10 kHz. Integrating the current results in the P-E_f loops in Fig. 4.3. An excellent agreement between simulation and experiment can be seen in Fig. 4.9a).

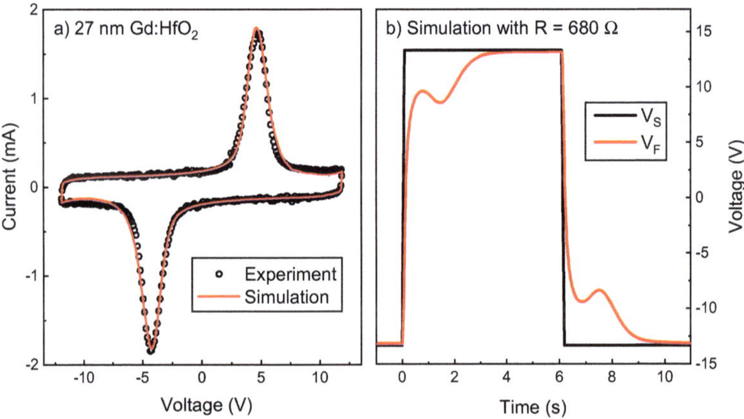

Figure 4.9: a) Modeled transient current response of a 27 nm Gd:HfO$_2$ capacitor to a 10 kHz triangular voltage excitation with $R = 0$. b) NC transient simulation of the same sample with series resistance $R = 680$ Ω. Copyright 2016 Wiley. Adapted with permission from Ref. [137].

In a next step, the resistance R was set to 680 Ω and the same waveform V_S as previously shown in Fig. 4.5 was applied in the simulation. The resulting NC transient V_F wafeform in Fig. 4.9b) shows a qualitatively similar behavior to the experimental data in Fig. 4.5. The used simulation parameters can be found in Table 4.1.

Table 4.1: Multigrain Landau-Khalatnikov simulation parameters used for obtaining the curves shown in Fig. 4.9. μ is the mean value and σ is the standard deviation when assuming a Gaussian distribution.

Parameter	μ	σ	Units
α	$-1.03 \cdot 10^9$	$4.8 \cdot 10^7$	m F^{-1}
β	$3.5 \cdot 10^{10}$	$1.23 \cdot 10^{10}$	m^5 F^{-1} C^{-2}
γ	0	0	m^9 F^{-1} C^{-4}
R_i	$1.45 \cdot 10^7$	$1.45 \cdot 10^6$	Ω
V_b	0.2	0.4	V
R	680	-	Ω
C_p	116	-	pF
A	33000	-	μm^2
t_f	27.34	-	nm
N	10000	-	grains
Δt	10	-	ns
I_{L0}	40	-	nA
V_{L0}	1	-	V

4.2 Transient negative capacitance in ferroelectric lead zirconate titanate

While HfO_2 based ferroelectrics are of great interest for applications, the experimental boundary conditions under which transient NC was observed in these materials was rather narrow, as shown in Section 4.1. Therefore, to investigate the underlying ferroelectric switching dynamics of transient NC effects, high-quality epitaxial samples were used, which enabled a larger variation of the experimental boundary conditions. A time-dependent multi-domain Ginzburg-Landau model was developed to model the NC switching transients and to get more insight into the domain nucleation and growth dynamics. This section is based on the results published in Ref. [218].

4.2.1 Electrical measurement setup

The general experimental setup shown in Fig. 4.10 is similar to the one used in Section 4.1 using an epitaxial PZT based ferroelectric capacitor instead of a polycrystalline HfO_2 based one.

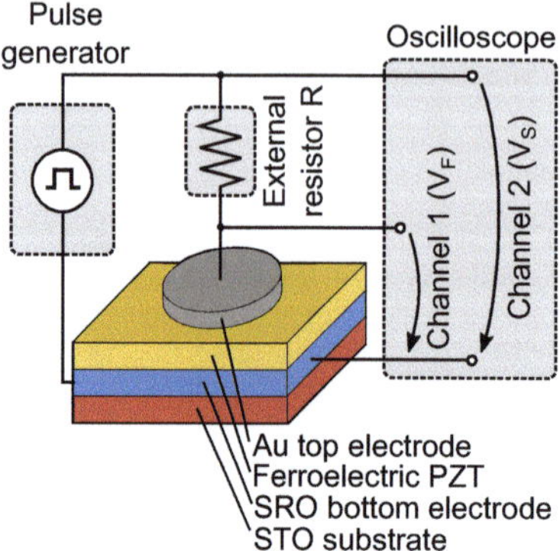

Figure 4.10: Schematic sample structure and measurement circuit applied to directly measure NC voltage transients in PZT via an oscilloscope. Reprinted from Ref. [218], with the permission of AIP Publishing.

The ferroelectric capacitor consists of a strontium titanate ($SrTiO_3$) substrate with a 20 nm thin metallic strontium ruthenium oxide ($SrRuO_3$) bottom electrode. A 100 nm thin epitaxial ferroelectric PZT layer of $Pb(Zr_{0.2}Ti_{0.8})O_3$ stoichiometry is covered by a Ti/Au top electrode as described in Table 3.1. Again, the voltage V_S, which was applied to the series connection of the capacitor and an external resistor as well as the voltage across the capacitor V_F was measured with a digital oscilloscope. Exemplary NC voltage transients for a capacitor area of $A = (50\ \mu m)^2$, $R = 3360\ \Omega$ and a pulse width of 25 µs can be seen in Fig. 4.11, with magnifications of the NC regions in Fig. 4.11b,c).

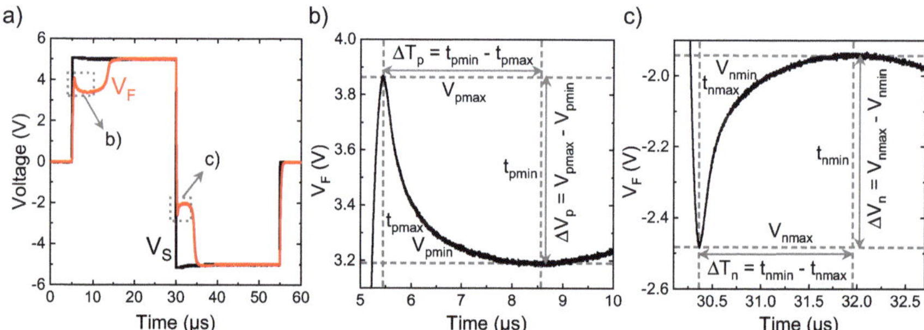

Figure 4.11: a) Voltage transients measured across the pulse generator V_S and the ferroelectric V_F for a capacitor with $A = (50\ \mu m)^2$ and $R = 3360\ \Omega$. Magnifications of the time frames where the ferroelectric capacitance is negative are shown in b) for positive V_S and in c) negative V_S. Reprinted from Ref. [218], with the permission of AIP Publishing.

The following characteristic parameters of the NC transients were defined: The NC transient times ΔT_p and ΔT_n as well as the NC voltage drop ΔV_p and ΔV_n, for the positive and negative applied pulses, respectively (see Fig. 4.11b,c)). ΔT_p is defined as the time span starting from the drop of V_F

at the time t_{pmax} until the subsequent increase of V_F at the time t_{pmin}. The NC voltage drop ΔV_p is defined as the difference between the maximum voltage V_{pmax} and the minimum voltage V_{pmin} of the NC transient. The quantities ΔT_n and ΔV_n for the reverse NC switching transient are defined analogously as shown in Fig. 4.11c) from the parameters t_{nmin}, t_{nmax}, V_{nmax} and V_{nmin}. By using Eq. 4.4, one can then calculate the ferroelectric charge Q_F on the capacitor as well as the ferroelectric polarization as $P \approx Q_F/A$. Furthermore, when calculating the electric field inside the ferroelectric as $E_f = V_F/t_f$, the P-E_f hysteresis loop is obtained as shown in Fig. 4.12.

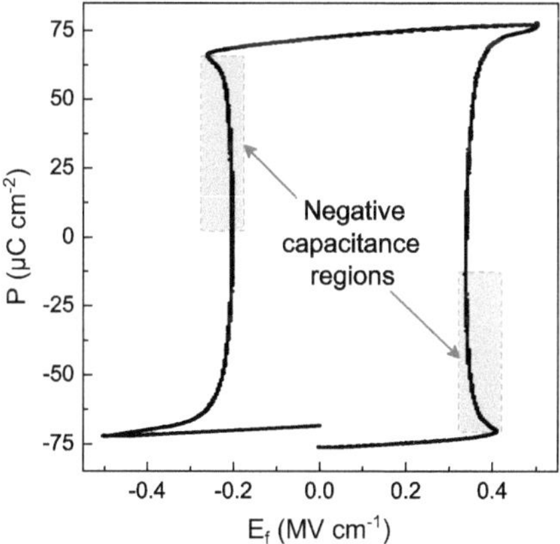

Figure 4.12: Polarization-electric field hysteresis for a capacitor with $A = (50\ \mu m)^2$ and $R = 3360\ \Omega$ calculated from NC voltage transients shown in Fig. 4.11. Reprinted from Ref. [218], with the permission of AIP Publishing.

For both switching directions around the coercive fields, the NC regions are present as regions with $dP/dE_f < 0$. However, in contrast to the homogeneous Landau 'S'-curve P-E_f model, a large hysteresis is present in Fig. 4.12. As previously discussed, the reason is the fast screening of polarization charge by the free electrons in the electrodes in MFM capacitor structures. The stability of the intrinsic NC state in similar structures will be thoroughly investigated in Chapter 5. However, it is intuitively clear that the small screening length of the metal electrodes leads to rather small depolarization fields by an almost ideal compensation of ferroelectric polarization charges at the electrode interfaces and the flow of screening charge can only be slowed down, but not stopped by the external resistance.

4.2.2 Multi-domain Ginzburg-Landau simulations

To better understand how the nucleation and growth of domains known from ferroelectric PZT films results in a NC voltage transient, a time-dependent Ginzburg-Landau approach was applied, based on the Gibb's free energy density

$$G = \alpha P^2 + \beta P^4 + \gamma P^6 - E_f P + k(\nabla P)^2, \tag{4.25}$$

where the ferroelectric background permittivity is set to zero (see Eq. 2.2). The time-dependence is again given by the Landau-Khalatnikov formalism in Eq. 2.21. Using a finite difference approach for spatial discretization, the Gibb's free energy density of each simulated cell i is given by

$$G_i = \alpha_i P_i^2 + \beta_i P_i^4 + \gamma_i P_i^6 - E_f P_i + k \sum_l \left(\frac{P_i - P_l}{\Delta x} \right)^2, \tag{4.26}$$

where the index l corresponds to the directly adjacent cells and Δx is the grid spacing between two cells, which is assumed identical for x- and y-directions. By substituting Eq. 4.26 into Eq. 2.21, one then obtains for a cell with four nearest neighbors

$$-\rho_i \frac{\partial P_i}{\partial t} = 2\alpha_i P_i + 4\beta_i P_i^3 + 6\gamma_i P_i^5 - E_f + \frac{8k}{\Delta x^2} P_i - \frac{2k}{\Delta x^2} \sum_l P_l. \tag{4.27}$$

Cells at the edge and at the corners of the grid have three and two nearest neighbors, respectively, corresponding to open boundary conditions. By using a time-discretization of Eq. 4.27, the evolution of P_i as a function of E_f can be calculated for each time step. The coupling to the external circuit is given by the same equations as discussed in Section 4.1. The variation of the ferroelectric parameters is modeled with Gaussian distributions for α, β, γ and E_{bias} as well as a uniform distribution of ρ between a minimum and maximum value of ρ_{min} and ρ_{max}, respectively. Figure 4.13a) shows a comparison between an experimental and a simulated NC voltage transient for positive applied V_S for the simulation parameters in Table 4.2.

Table 4.2: Parameters fitted from the experimental data, which were used for time-dependent Ginzburg-Landau simulations. α, β, γ, and E_{bias} are normally distributed and ρ is uniformly distributed. Reprinted from Ref. [218], with the permission of AIP Publishing.

Parameter	μ	σ	Units
α	$-4.4 \cdot 10^7$	$3.3 \cdot 10^6$	$\mathrm{m\ F^{-1}}$
β	$-9.1 \cdot 10^6$	$6.8 \cdot 10^5$	$\mathrm{m^5\ F^{-1}\ C^{-2}}$
γ	$7.4 \cdot 10^7$	$5.6 \cdot 10^6$	$\mathrm{m^9\ F^{-1}\ C^{-4}}$
E_{bias}	$7.2 \cdot 10^6$	$5.4 \cdot 10^5$	$\mathrm{V\ m^{-1}}$

Parameter	Value	Units
k	$2.5 \cdot 10^{-7}$	$\mathrm{m^3\ F^{-1}}$
ρ_{min}	0.4	$\Omega\,\mathrm{m}$
ρ_{max}	13	$\Omega\,\mathrm{m}$

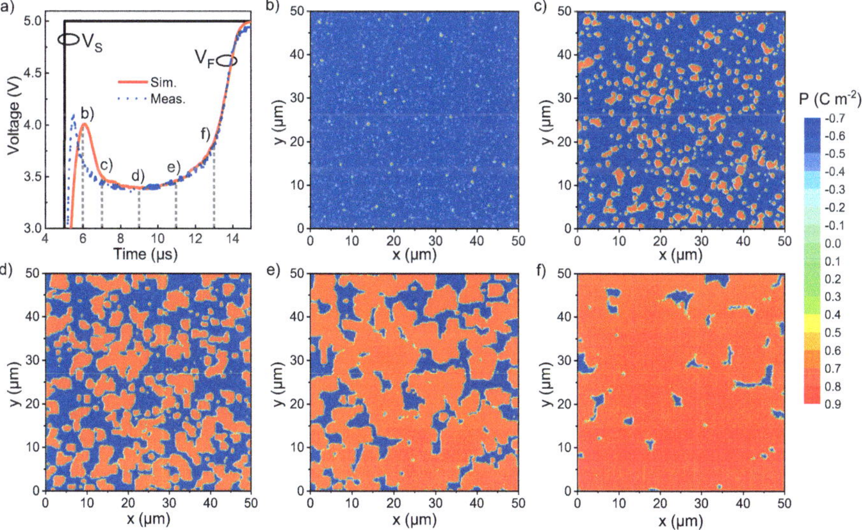

Figure 4.13: a) Experimental and simulated voltage transients across the ferroelectric V_F in response to the applied voltage V_S. b)–f) Simulated spatial polarization distribution in the ferroelectric during switching for b) $t = 6$ µs, c) 7 µs, d) 9 µs, e) 11 µs, and f) 13 µs, respectively. Reprinted from Ref. [218], with the permission of AIP Publishing.

The spatial variation of P across the capacitor area for different simulation time steps is shown in Fig. 4.13b)-f). Only in the initial switching stage ($t < 7$ µs), there is a quantitative disagreement between experiment and simulation, which indicates that the nucleation of new domains is slightly faster than this simplified time-dependent Ginzburg-Landau model predicts (see Fig. 4.13b)). Nevertheless, the agreement of simulation and experiment is excellent after this initial nucleation period, which suggests that the simulated domain dynamics Fig. 4.13c)-f) might be a good approximation of the experimental behavior.

Interestingly, the NC region of the voltage transient seems to only persist as long as domains are nucleated and expand unimpeded throughout the sample. However, once the domains start to coalesce, V_F increases again, corresponding to the positive capacitance regime. Therefore, it seems that the persistence of the transient NC state is mainly limited by the speed of lateral domain wall motion. The slower the domain walls move, the longer the NC voltage transient. The behavior in Fig. 4.13 also explains why a large hysteresis is always expected in these kinds of MFM structures: For the transient NC effect to be observed, most of the ferroelectric has to be first switched to one direction, e.g. up. Then when switching the polarization to the opposite direction (i.e. down), NC can be measured until roughly half of the domains have been switched. However, to observe the same transient NC effect again, the MFM has to be reset to the same initial condition. This process involves a hysteresis since the nucleation and growth dynamics which govern these switching processes are inherently dissipative [219].

4.2.3 Analytical model for transient negative capacitance

To further investigate the NC domain dynamics, the effect of different voltage pulse levels V_{max}, capacitor areas A and external resistances R was analyzed. Especially, the NC transient times $\Delta T_{p,n}$ were of interest in these dynamic switching measurements. As can be already seen in Fig. 4.11, $\Delta T_p > \Delta T_n$ in these samples, which would not be expected for perfectly symmetric MFM capacitors. However, it is also clear from Fig. 4.12 that the P-E_f hysteresis is shifted along the E-axis towards higher electric fields. This shows that there is an internal bias field E_{bias}, which can originate e.g. from different work function electrodes or asymmetric charge distributions inside the capacitor [220, 221]. From Fig. 4.12 and in consistency with the Ginzburg-Landau simulations, an $E_{bias} = 72 \text{ kV cm}^{-1}$ was extracted.

In Fig. 4.14, the NC transient times are shown as a function of the inverse of V_{max} in a semi-logarithmic plot. In this depiction, the dependency appears approximately linear. While this is true for both ΔT_p and ΔT_n, the asymmetry due to E_{bias} is obvious. However, this offset due to E_{bias} can easily be corrected by plotting ΔT_p against $1/(V_{max} - t_f E_{bias})$ and ΔT_n against $1/(V_{max} + t_f E_{bias})$, which is shown by the blue data points (denoted ΔT) in Fig. 4.14. As expected, all data points now lie on the same trend line.

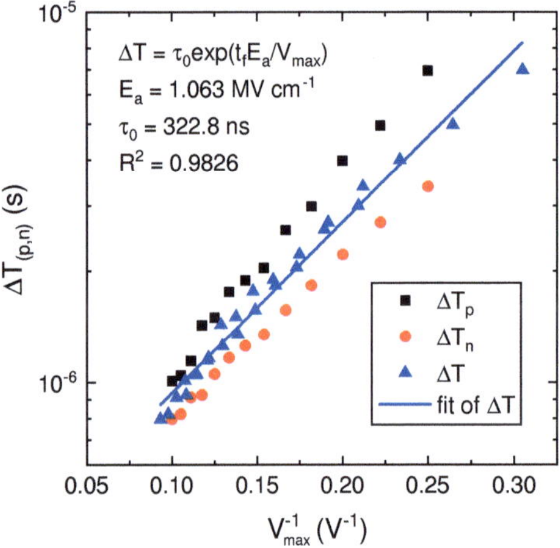

Figure 4.14: NC transient times with (ΔT) and without correction ($\Delta T_{p,n}$) for the internal bias field as a function of V_{max} with $R = 10 \text{ k}\Omega$ and $A = (35 \text{ μm})^2$. The solid line shows the analytical fit of ΔT. Reprinted from Ref. [218], with the permission of AIP Publishing.

Such an electric field dependence of the switching time has been known for a long time as the empirical Merz's law [57], which can be written as

$$\frac{1}{v_{DW}} \propto \tau_{sw} = \tau_0 \exp\left(\frac{E_a}{E_f}\right)^m,$$ (4.28)

where v_{DW} is the domain wall velocity, τ_{sw} is the characteristic switching time, τ_0 is the intrinsic switching time constant, E_a is the activation field and m is the exponent. By fitting this model to the ΔT data in Fig. 4.14, the following parameters are obtained: $m = 1$, $E_a = 1.063$ MV cm^{-1} and $\tau_0 = 322.8$ ns. An exponent $m = 1$ signifies that the polarization switching is dominated by a domain wall creep process [222, 223]. This is in accordance with previous results for epitaxial PZT thin films, where a similar activation field of 1 MV cm^{-1} was reported [222, 224].

By combining simple electric circuit equations with Merz's law, an analytical model for the NC transient time as a function of the external parameters can be derived. Therefore, when assuming a homogeneous ferroelectric layer for simplicity, the applied voltage can be approximated as

$$V_S \approx \frac{dP}{dt}(RA + t_f \rho) + (2\alpha P + 4\beta P^3 + 6\gamma P^5)t_f,$$ (4.29)

where the first two terms on the right hand side are the voltage drops across the external resistor R and the internal resistance ρ, respectively. This expression can be rearranged into

$$\frac{dP}{dt} \approx \frac{V_S - (2\alpha P + 4\beta P^3 + 6\gamma P^5)t_f}{RA + t_f \rho}.$$ (4.30)

From Eq. 4.30 it is clear that the change of P with time is limited by the external and internal resistances through the factor $(RA + t_f \rho)$, i.e. the larger the sum of these resistances, the slower the change in P. On the other hand, this means that for a fixed change in polarization ΔP, the switching time Δt will be proportional to $(RA + t_f \rho)$. Therefore, a combined analytic formula for the NC transient time is proposed in the form of

$$\Delta T = |C_f|(RA + t_f \rho)\exp\left(\frac{E_a}{E_f}\right),$$ (4.31)

where C_f is the average NC per area and ρ is the average internal resistance of the ferroelectric. The activation field E_a was already extracted from the fit of the electric field-dependent data in Fig. 4.14. The 'missing' parameters C_f and ρ can be extracted by changing either R or A while keeping the applied electric field E_f constant. For these experiments, ΔT can be calculated from $\Delta T_{p,n}$ by using

$$\Delta T = \Delta T_{p,n}\exp\left[E_a\left(\frac{1}{E_f} - \frac{1}{E_f \pm E_{bias}}\right)\right].$$ (4.32)

The resulting NC transient times as a function of the capacitor area for $R = 10$ kΩ and $V_{max} = 5$ V are shown in Fig. 4.15. As expected from Eq. 4.32, the dependence of ΔT on the area is linear, which was fitted by a simple $\Delta T = c_A A + \tau_A$ expression as depicted in Fig. 4.15.

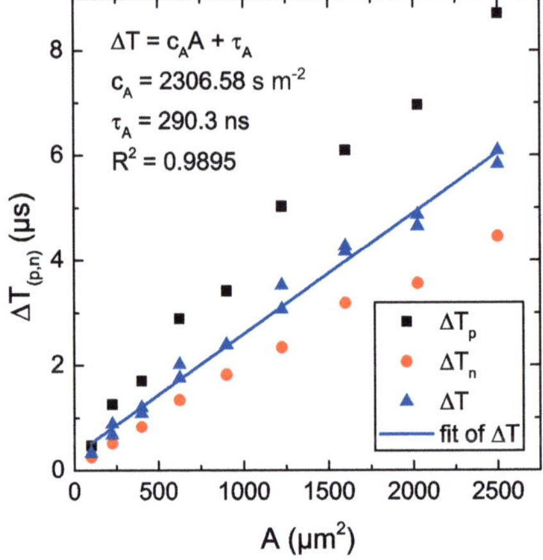

Figure 4.15: NC transient times with (ΔT) and without correction ($\Delta T_{p,n}$) for the internal bias field as a function of A with $R = 10$ kΩ and $V_{max} = 5$ V. The solid line shows the linear fit of (ΔT). Reprinted from Ref. [218], with the permission of AIP Publishing.

From these fitted values $c_A = 2306.58$ s m^{-2} and $\tau_A = 290.3$ ns, the average NC of the ferroelectric C_f can be calculated as

$$|C_f| = \frac{c_A}{R}\exp\left(-\frac{E_a}{E_f}\right),\tag{4.33}$$

which yields $C_f = -2.3$ µF cm^{-2}. Furthermore, ρ can be estimated based on the fitted data from Fig. 4.15 as

$$\rho = \frac{\tau_A}{t_f|C_f|}\exp\left(-\frac{E_a}{E_f}\right),\tag{4.34}$$

thus resulting in $\rho \approx 15$ Ωm. However, since the fit in Fig. 4.15 might not be very accurate for $A \to 0$, a change of R should give a more reliable estimation of ρ. Therefore, the external resistance R was changed while keeping the area constant at 2500 µm^2 and $V_{max} = 5$ V. The dependence of ΔT and $\Delta T_{p,n}$ on R is shown in Fig. 4.16.

While the E_a and C_f values were taken from the fitting in Fig. 4.14 and Fig. 4.15, respectively, ρ was now fitted to the data in Fig. 4.16. It is apparent that ρ is different for positive and negative applied voltage pulses. Therefore, two different values of ρ were fitted for positive and negative NC

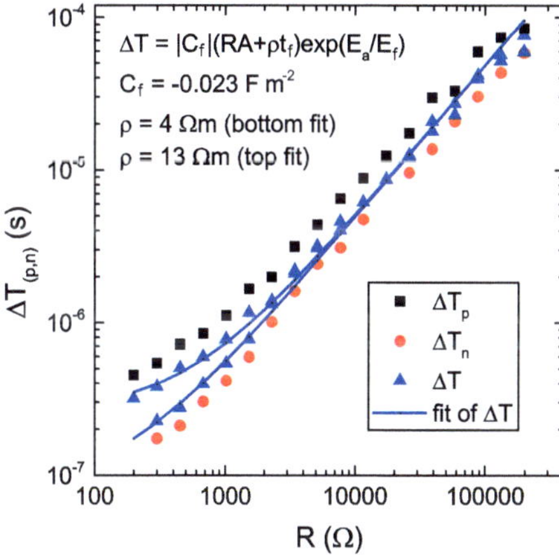

Figure 4.16: NC transient times with (ΔT) and without correction ($\Delta T_{p,n}$) for the internal bias field as a function of R with $A = (50\ \mu m)^2$ and $V_{max} = 5$ V. The solid lines show the linear fit of ΔT. Reprinted from Ref. [218], with the permission of AIP Publishing.

transients with 13 Ωm and 4 Ωm, respectively, which is consistent with the results of the Ginzburg-Landau simulations in Table 4.2. In the following section, the same PZT capacitor as characterized here was connected in series with a dielectric capacitor instead of a resistor to experimentally investigate the possibility of voltage amplification from transient NC.

4.3 Differential voltage amplification from transient negative capacitance

The idea to obtain passive differential voltage amplification from NC goes back to the first publication on NC by Salahuddin and Datta [4]. It was predicted that when applying a voltage V_S to a series connection of a positive capacitor and a negative capacitor, the voltage V_D across the former will be amplified with respect to the applied voltage V_S, see also Fig. 4.1. When considering the change of the voltages with time, the circuit equation can be rearranged into

$$\frac{dV_D}{dt} = \frac{dV_S}{dt} - \frac{1}{C_F}\frac{dQ}{dt},$$ (4.35)

where Q is the charge on the positive capacitor. Dividing Eq. 4.35 by dV_S/dt, the voltage amplification A_V can be written as

$$A_V = \frac{dV_D}{dV_S} = 1 - \frac{1}{C_F}\frac{dQ}{dV_S}.$$

(4.36)

Since dQ/dV_S is always positive, having a positive ferroelectric capacitance $C_F > 0$ would always result in $A_V < 1$, i.e. no differential voltage amplification. However, if $C_F < 0$, then $A_V > 1$. In the following, a method is described to directly measure A_V in a series circuit of a regular dielectric capacitor and the same ferroelectric PZT capacitor used in Section 4.2. The following section is based on the results published in Ref. [225].

4.3.1 Experimental measurement setup

Fig. 4.17 schematically shows the experimental setup. Initially, the ferroelectric is poled in one of its remanent polarization states as indicated in 4.17a). Then, a triangular voltage signal V_S is applied to the series connection of the ferroelectric and dielectric capacitor, while the voltage across the dielectric capacitor V_D is measured with an oscilloscope. As shown in the previous sections, during switching of the ferroelectric capacitor, the voltage V_F across it will be reduced and thus lead to a stronger increase of V_D and thus should enable $A_V > 1$ according to Eq. 4.36.

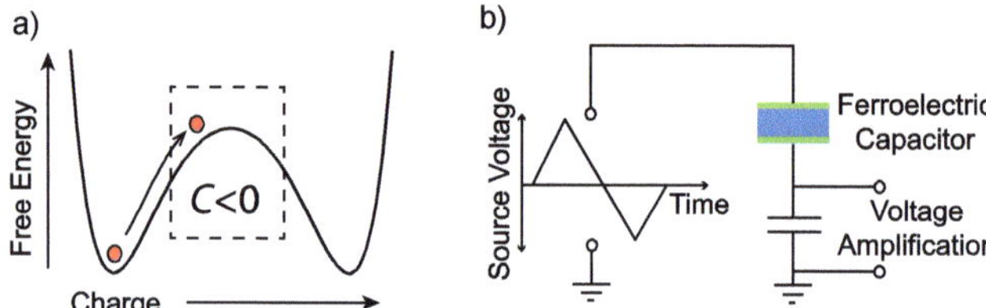

Figure 4.17: Voltage amplification due to ferroelectric NC: a) Energy landscape of a ferroelectric capacitor. The capacitance, C, is negative in the region enclosed by the dashed box. b) The experimental setup. V_S and V_D are the source voltage and the voltage across the dielectric capacitor, respectively. Reprinted from Ref. [225], with the permission of AIP Publishing.

For all experiments, 100 nm thick PZT capacitors with an area $A = (35\ \mu m)^2$ as described in Section 4.2 were used, while different dielectric capacitors ($C_D = 77...440$ pF) were connected in series.

4.3.2 Differential voltage amplification

Fig. 4.18 shows the measurement results for a triangular voltage signal with 50 µs period and 10 V maximum amplitude for a dielectric capacitor with $C_D = 440$ pF. By comparing the slope dV_S/dt of the applied voltage to the measured slope dV_D/dt as depicted in Fig. 4.18a,b), one can immediately see that V_D indeed changes faster than V_S during a certain time frame in which the ferroelectric is switching. Looking at Eq. 4.35, such a behavior would be impossible if C_F was positive in these time frames. Indeed, one can directly calculate the measured A_V as a function of V_S, which is shown in Fig. 4.18c).

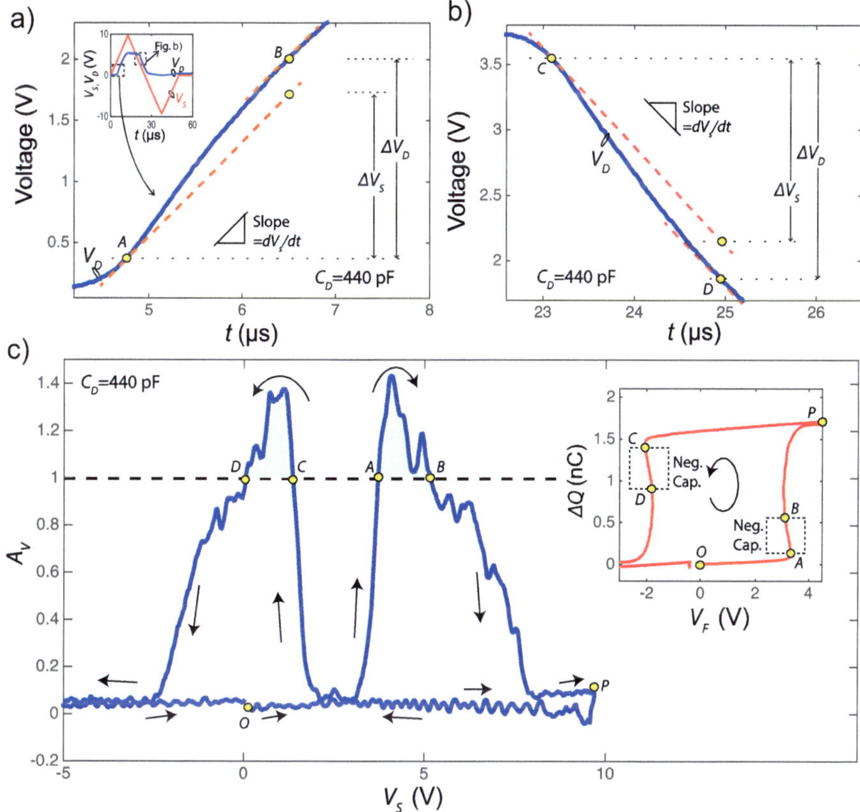

Figure 4.18: Voltage amplification in a ferroelectric-dielectric series circuit. a) and b) waveforms corresponding to the voltage across the positive capacitor V_D in response to a bipolar triangular voltage pulse V_S: 0 V $\rightarrow$ +10 V $\rightarrow$ -10 V $\rightarrow$ 0 V with period $T = 50$ μs during 4.2 μs $< t <$ 6.7 μs (a)) and 22.6 μs $< t <$ 25.3 μs (b)). $C_D =$ 440 pF. The dashed red lines have the same slew rates as those of the $V_S(t)$-t curves in these time frames (i.e., they are | | to the V_S-t curves). The inset in a) shows the waveforms corresponding to the source voltage V_S and V_D during the entire cycle. Differential amplification is observed in the regions corresponding to the green shades, segments AB and CD in a) and b), respectively. c) Amplification A_V (= dV_D/dV_S) as a function of V_S. The inset in c) shows ferroelectric charge (ΔQ)-voltage (V_F) characteristics extracted from the waveforms. Reprinted from Ref. [225], with the permission of AIP Publishing.

The A_V vs. V_S curve has a butterfly hysteresis shape qualitatively similar to a C-V measurement. As previously discussed in Section 4.2, on the onset of domain nucleation and unrestricted growth, the ferroelectric exhibits transient NC resulting in $A_V > 1$, see also the ΔQ vs. V_F curve in the inset of Fig. 4.18c). Only between the points A and B as well as C and D, $C_F < 0$. Everywhere else,

$C_F > 0$ resulting in no voltage amplification ($A_V < 1$). Since the ferroelectric P-E_f (or ΔQ-V_F) curve extracted here is similar to the one in Fig. 4.12, the physical origin of NC in this kind of experiment should be the same as in the ferroelectric capacitor-resistor series circuit as discussed before. In a subsequent experiment, the capacitance of the dielectric capacitor C_D was changed, while keeping the applied voltage waveform V_S and the ferroelectric capacitor the same. The A_V vs. V_S graphs for $C_D = 240$ pF and 145 pF are shown in Fig. 4.19a) and b), respectively.

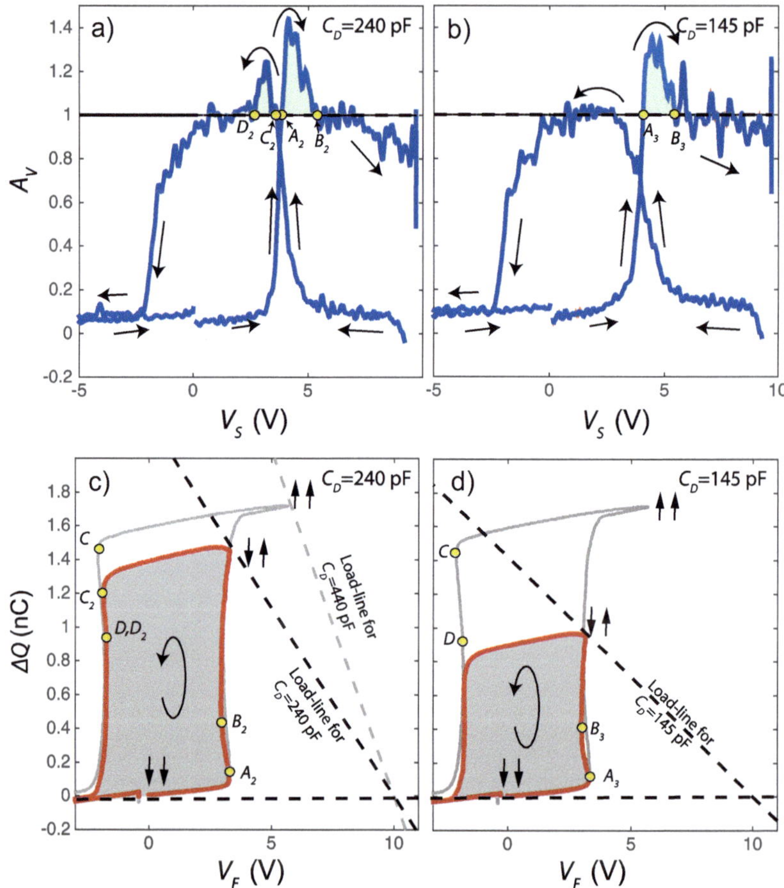

Figure 4.19: Effect of dielectric capacitance on the voltage amplification. a) and b) shows the amplification A_V as a function of V_S for $C_D = 240$ pF (a)) and 145 pF (b)). c) and d) show the extracted charge-voltage characteristics of the ferroelectric capacitor for $C_D = 240$ pF (c)) and 145 pF (d)). These loops are overlaid on the ferroelectric charge-voltage characteristics extracted for $C_D = 440$ pF, which is also shown in the inset of Fig. 2c). The load lines for the dielectric capacitor $\Delta Q = C_D(V_S - V_F)$ corresponding to $V_S = +10$ V are plotted for $C_D = 240$ and 145 pF in (c)) and (d)), respectively. The load-line ($V_S = +10$ V) for $C_D = 440$ pF is plotted in both of them. Reprinted from Ref. [225], with the permission of AIP Publishing.

It can be seen that with decreasing C_D, while the voltage amplification in the positive V_F region is unchanged, the amplification maximum during reverse switching ($V_F < 0$) is reduced and even vanishes for $C_D = 145$ pF with $A_V \approx 1$. The reason for this behavior can be understood by applying the load-line technique and looking at the ΔQ-V_F curves in Fig. 4.12c) and d) for $C_D = 240$ pF and 145 pF, respectively. For comparison, the ΔQ-V_F hysteresis for $C_D = 440$ pF is shown in gray. The black dashed lines show the load-lines $\Delta Q = C_D(V_S - V_F)$ of the respective dielectric capacitors for the maximum applied voltage of $V_S = 10$ V.

The intersection between the load-line and the ferroelectric ΔQ-V_F curve determines the operating point of the circuit. Since the slope of the load-line scales with C_D, the operating point for $V_S = 10$ V shifts to much lower ΔQ for smaller C_D. In contrast to the $C_D = 440$ pF case, where the whole ferroelectric polarization can be switched, the reduced slope of the load-line leads to sub-loop operation for lower C_D, which means that some domains in the material do not switch because the maximum ΔQ is given by C_D and the maximum of V_S. Therefore, when only a fraction of the domains are switched for positive V_F, the transient NC effect from nucleation and unrestricted domain growth as discussed in Section 4.2 will be reduced and eventually vanish in the case where only domain coalescence happens during reverse switching (when $V_F < 0$). This shows that the simple picture of single-domain Landau stabilization of NC in Fig. 2.11 is not accurate for macroscopic MFM capacitors, where domains can easily form.

4.3.3 Landau-Khalatnikov simulations

Lastly, the qualitative circuit behavior was simulated dynamically using a homogeneous Landau-Khalatnikov approach, where the voltage across the ferroelectric is defined as

$$V_F = R_i \frac{\mathrm{d}Q_F}{\mathrm{d}t} + (aQ_F + bQ_F^3) + V_b. \tag{4.37}$$

Leakage currents through the ferroelectric and dielectric capacitors were modeled by a parallel resistance R_L and R, respectively (see Fig. 4.20a)). Therefore, the following circuit equation must be solved self-consistently with Eq. 4.37:

$$\frac{\mathrm{d}Q_F}{\mathrm{d}t} + \frac{V_F}{R_L} = C_D \frac{\mathrm{d}V_D}{\mathrm{d}t} + \frac{V_D}{R}. \tag{4.38}$$

The parameters used for the simulation were $A = 1225$ µm^2, $t_f = 100$ nm, $V_b = 1.5$ V, $C_D = 440$ pF, $\alpha = -5 \cdot 10^7$ m F^{-1} and $\beta = -4.5 \cdot 10^7$ m^5 F^{-1} C^{-2}. The resistances R and R_L were set to ~ 8 MΩ. The resulting dynamic response of the voltage across the dielectric capacitor V_D to the triangular input voltage V_S as a function of time is shown in Fig. 4.20b). Similar as observed in the experiment, between the points A and B as well as C and D, a voltage amplification $A_V > 1$

can be seen in Fig. 4.20c). The qualitatively similar behavior in the simplified Landau based simulation supports the interpretation of NC as the origin of the differential voltage amplification. The extracted P-E_f hysteresis loop in the inset of Fig. 4.20c) furthermore coincides with results from previous simulations of a resistor in series with a ferroelectric capacitor [127].

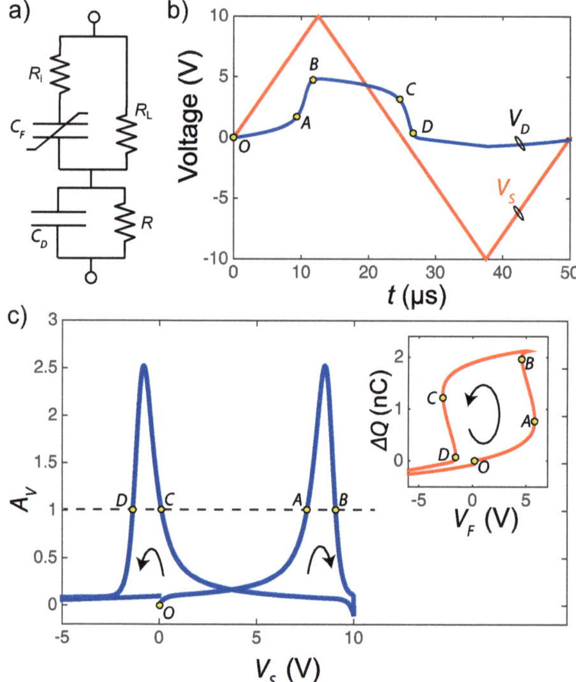

Figure 4.20: Simulation results. a) Circuit diagram of the simulation. C_F, R_i, and R_L represent the capacitance, the internal resistance, and the leakage resistance of the ferroelectric capacitor. C_D and R represent the capacitance and the leakage resistance of the dielectric capacitance. b) Simulated waveforms corresponding to V_S and V_D of the circuit shown in a) in response to a bipolar triangular pulse V_S: 0 V $\rightarrow$ +10 V $\rightarrow$ -10 V $\rightarrow$ 0 V with period $T = 50$ μs. Amplification is observed in the segments, AB and CD. c) Simulated amplification A_V as a function of V_S. $A_V \geq 1$ in the segments, AB and CD. The inset shows the ferroelectric charge-voltage characteristics extracted from the waveforms in b). Reprinted from Ref. [225], with the permission of AIP Publishing.

4.4 Summary

Transient NC in ferroelectrics can be directly observed by applying large enough voltage signals to a series connection of a metal-ferroelectric-metal capacitor with either a resistor or a dielectric capacitor. The NC can be inferred from the reduction of the ferroelectric voltage while the ferroelectric charge is increasing and vice versa. These transient NC effects are only observed for a limited time after the onset of ferroelectric switching and result in a significant polarization hysteresis, which is undesirable for most NC applications. In HfO_2 based polycrystalline thin film capacitors, the transient NC effect was observed only for relatively small series resistors and could be modeled by a multi-grain Landau-Khalatnikov approach. In epitaxial ferroelectric PZT capacitors, on the other hand, transient NC was observed for a wide range of series resistances, capacitor areas and applied voltage amplitudes. A time-dependent Ginzburg-Landau model was applied to simulate the NC domain dynamics during ferroelectric switching. These numerical simulations indicate that the observation of transient NC corresponds to a region of domain nucleation and unimpeded growth, before domains start to coalesce. Typical domain wall creep motion was identified as the decisive factor limiting the transient NC switching speed. An analytical model for the NC transient time was developed. Lastly, it was shown that when connecting similar epitaxial PZT capacitors in series with a dielectric capacitor, a differential voltage amplification due to transient NC can be measured during ferroelectric switching. The stabilization of NC in a series connection of macroscopic ferroeletric and dielectric capacitors is not possible due to the fast screening of polarization charges by the metal electrodes and the formation of domains. Therefore, in the following, the conditions for obtaining a stabilized intrinsic NC will be theoretically investigated.

5 Stabilizing intrinsic ferroelectric negative capacitance

While the direct experimental observation of transient NC in metal-ferroelectric-metal capacitors as described in the previous chapter is intriguing, the prevalent polarization hysteresis is undesirable for most NC applications. Therefore, in this chapter, the possible stabilization of intrinsic NC will be investigated theoretically, considering the possibility of domain formation and different proposed device designs. In general, the two different proposed structures for NC stabilization are the metal-ferroelectric-metal-insulator-metal (MFMIM) and metal-ferroelectric-insulator-metal (MFIM) structures, see Fig. 5.1. While for the application of NC in a transistor, a semiconducting material is used to stabilize the NC, the resulting non-linear positive capacitance complicates analytical solutions. Therefore, to better understand the basic physics of NC stabilization, the case of a linear stabilizing dielectric layer is investigated here.

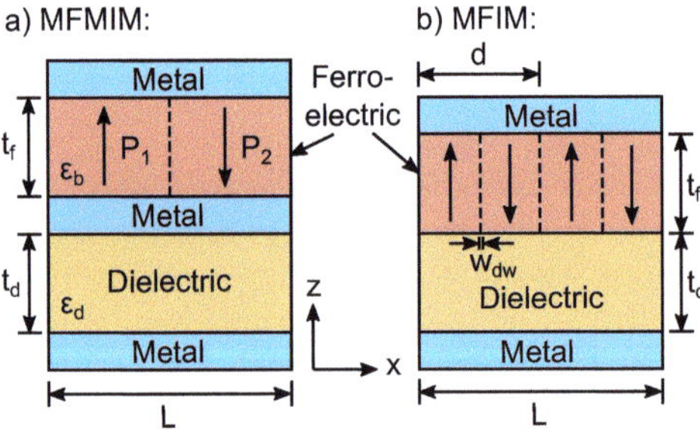

Figure 5.1: Two different suggested structures for stabilization of NC: a) MFMIM and b) MFIM stacks. Dashed lines indicate domain walls with width w_{dw}. Arrows indicate polarization directions. t_f, t_d, ε_b, ε_d, d and L are the ferroelectric thickness, dielectric thickness, ferroelectric background permittivity, dielectric permittivity, domain period and lateral capacitor size, respectively. Adapted from Ref. [135] with permission from The Royal Society of Chemistry.

While most early studies have only considered a homogeneous single-domain ferroelectric [226], the presence of large depolarization fields is known to facilitate domain formation. Therefore, considering the effects of domain formation on intrinsic NC stabilization are of great practical importance. For simplicity, here, only out-of-plane (z-axis) spontaneous polarization in the ferroelectric as well as symmetric stripe-domains are considered. Domain walls are assumed to be abrupt and thin compared to the domain width (so-called Kittel domains, see Fig. 5.1). Furthermore, domain wall motions are neglected, which is why only intrinsic NC effects are discussed in the following. This chapter is based on the results published in Refs. [135, 227, 228].

5.1 Helmholtz free energy potentials

First, the Helmholtz free energy density F_f including the domain wall energy and electrostatic self-energy of the ferroelectric is considered as

$$F_f = \alpha P_S^2 + \beta P_S^4 + \gamma P_S^6 + k(\nabla P_S)^2 + \frac{\varepsilon_0 \varepsilon_b}{2} E_f^2, \tag{5.1}$$

where k is the domain wall coupling factor. For further simplification, a second-order phase transition is assumed, i.e. $\alpha < 0$, $\beta > 0$ and $\gamma = 0$. The lateral dimension of the capacitor structures is L and the domain width is $d/2$ as indicated in Fig. 5.1. Furthermore, the spontaneous polarization in the left domain is called P_1 and in the right domain P_2. The energy density due to the ferroelectric anisotropy including both domains is then given by

$$F_a = \frac{1}{2}(\alpha P_1^2 + \beta P_1^4) + \frac{1}{2}(\alpha P_2^2 + \beta P_2^4) = \frac{\alpha}{2}(P_1^2 + P_2^2) + \frac{\beta}{2}(P_1^4 + P_2^4). \tag{5.2}$$

To visualize Eq. 5.2, a 3D plot of F_a as a function of both P_1 and P_2 is shown in Fig. 5.2.

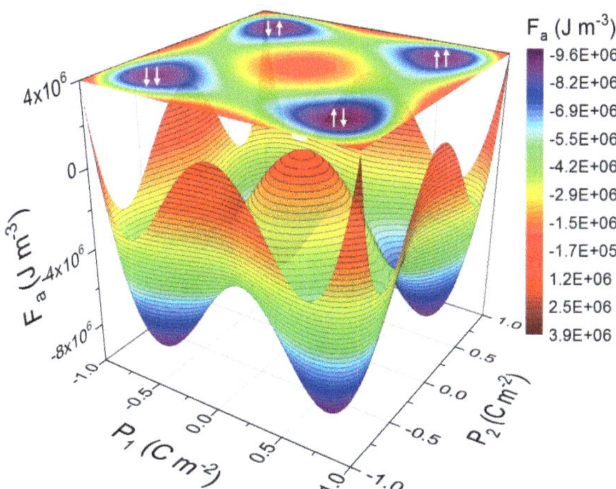

Figure 5.2: Free anisotropy energy of a ferroelectric with two equally sized domains as a function of the polarization of two domains of equal size P_1 and P_2. Arrows indicate polarization directions. Adapted from Ref. [135] with permission from The Royal Society of Chemistry.

Four degenerate energy minima can be observed, two of which correspond to the up- and down pointing polarization ($\uparrow\uparrow$ or $\downarrow\downarrow$) in the single-domain case ($P_1 = P_2$) and two corresponding to antiparallel ($\uparrow\downarrow$ or $\downarrow\uparrow$) domain formation ($P_1 = -P_2$). In reality, the antiparallel domain minima would be slightly higher in energy compared to the homogeneous case, due to the domain wall energy contribution neglected in Fig. 5.2. The total domain wall energy contribution F_{DW} can be calculated by multiplying the energy density of a single domain wall (per domain wall area) $F_{single,DW}$ with the number of domain walls in the sample N_{DW} and then normalizing this to the length of the sample L to obtain the energy density per unit volume:

$$F_{DW} = \frac{F_{single,DW} N_{DW}}{L}.$$ (5.3)

The energy density of a single domain wall is given by

$$F_{single,DW} = w_{dw} k (\nabla P)^2 = w_{dw} k \frac{(P_1 - P_2)^2}{w_{dw}^2} = k \frac{(P_1 - P_2)^2}{w_{dw}}.$$ (5.4)

The total number of domain walls can be written as

$$N_{DW} = \frac{2L}{d} - 1.$$ (5.5)

By combining Eq. 5.3 and 5.5 the domain wall energy density per volume is given by

$$F_{DW} = \frac{1}{L} \left(\frac{2L}{d} - 1 \right) \frac{k}{w_{dw}} (P_1 - P_2)^2 = \left(\frac{2}{d} - \frac{1}{L} \right) \frac{k}{w_{dw}} (P_1 - P_2)^2.$$ (5.6)

The total free energy density of the ferroelectric is therefore given by $F_f = F_a + F_{DW} + \varepsilon_0 \varepsilon_b E_f^2 / 2$, which yields the final expression

$$F_f = \frac{\alpha}{2}(P_1^2 + P_2^2) + \frac{\beta}{2}(P_1^4 + P_2^4) + \left(\frac{2}{d} - \frac{1}{L} \right) \frac{k}{w_{dw}} (P_1 - P_2)^2 + \frac{\varepsilon_0 \varepsilon_b}{2} E_f^2.$$ (5.7)

Similarly to Eq. 5.7, the free energy density of the isotropic, linear dielectric layer can be written as

$$F_d = \frac{\varepsilon_0 \varepsilon_d}{2} E_d^2,$$ (5.8)

where E_d is the electric field inside the dielectric. The electric displacement field can then be calculated as $D_d = dF_d/dE_d = \varepsilon_0 \varepsilon_d E_d$ and the permittivity is given by $(d^2 F_d/dD_d^2)^{-1} = \varepsilon_0 \varepsilon_d$ [10]. For the following investigation of intrinsic NC stabilization, Eq. 5.7 and 5.8 together with the corresponding electrostatic boundary conditions are applied. To assess the stability of the NC state at $P_{1,2} \approx 0$, it is sufficient to investigate the stability of the single-domain state ($P_1 = P_2$) and of the antiparallel domain state ($P_1 = -P_2$), which are orthogonal in the (P_1, P_2)-plane.

5.2 Single-domain case

In the single-domain case $P_1 = P_2 = P_S$, when leakage currents through the ferroelectric and dielectric are neglected and there are no free charges in the materials or at their interfaces, both MFMIM and MFIM stacks should be governed by the same electrostatics. Therefore, in this section no distinction is made between the two structures. Furthermore, the electrostatics can be solved in the z-direction only, significantly simplifying the equations. The first electrostatic boundary condition is the continuity of the normal component of the electric displacement field as

$$\varepsilon_0 \varepsilon_d E_d = \varepsilon_0 \varepsilon_b E_f + P_S. \tag{5.9}$$

Furthermore, the sum of all voltages across the stack must be zero, which yields

$$V = t_f E_f + t_d E_d. \tag{5.10}$$

By inserting Eq. 5.9 into Eq. 5.10, the electric fields inside the ferroelectric and dielectric are given by

$$E_f = \frac{1}{t_f C_0}(-P_S + C_d V), \tag{5.11}$$

and

$$E_d = \frac{1}{t_d C_0}(P_S + C_{f,b} V), \tag{5.12}$$

respectively. The capacitance C_0 is defined as the sum of the dielectric capacitance C_d and the ferroelectric background capacitance $C_{f,b}$ as

$$C_0 = C_d + C_{f,b} = \varepsilon_0 \left(\frac{\varepsilon_d}{t_d} + \frac{\varepsilon_b}{t_f} \right). \tag{5.13}$$

Under the short-circuit condition ($V = 0$) and using Eq. 5.7 and 5.11, the free energy density in the ferroelectric is then given by

$$F_f(P_1 = P_2 = P_S) = \left(\alpha + \frac{C_{f,b}}{2 t_f C_0^2} \right) P_S^2 + \beta P_S^4. \tag{5.14}$$

Analogously, the free energy density in the dielectric can be obtained from Eq. 5.8 and 5.12, yielding

$$F_d = \frac{C_d}{2t_d C_0^2} P_S^2.$$

(5.15)

The total energy density per unit area can the be written as

$$F = t_f F_f + t_d F_d = \left(\alpha t_f + \frac{1}{2C_0} \right) P_S^2 + \beta t_f P_S^4.$$

(5.16)

From Eq. 5.16 the NC stability condition at $P_S = 0$ can be immediately seen, since the second derivative of F with respect to P_S must be positive. Otherwise, the NC state at $P_S = 0$ would be unstable. This means that $2\alpha t_f + 1/C_0 > 0$ must be fulfilled, where the term proportional to P_S^4 can be neglected for small P_S. By rearranging this condition, an expression for the critical ferroelectric thickness can be obtained as

$$t_f < t_{f,max} = -\frac{1}{2\alpha C_d}(1 + 2\varepsilon_0 \varepsilon_b \alpha).$$

(5.17)

The result in Eq. 5.17 is similar to the original result suggested by Salahuddin and Datta [4]. However, there is an additional term related to the background permittivity of the ferroelectric, which was not considered before. This shows, that the critical thickness for NC in the single-domain state is reduced by the (positive) background permittivity of the ferroelectric. Furthermore, when looking at Eq. 5.17, another condition can be derived when $t_{f,max} > 0$ must be satisfied, which results in

$$\varepsilon_b \alpha > -\frac{1}{2\varepsilon_0}$$

(5.18)

The condition in Eq. 5.18 means that if the ferroelectric background permittivity is too large compared to the α parameter, the NC state cannot be stabilized at all.

5.3 Multi-domain case

While the conditions for intrinsic NC stabilization in the single-domain case are rather simple, additionally the stability against domain formation must be ensured. Therefore, in the following the antiparallel $P_1 = -P_2 = P_S$ case is considered. However, in this case, the electrostatics in the MFMIM and MFIM case are different, which is why they have to be considered separately. Since the electrostatics of the MFMIM structure can be again solved in z-direction only, this case will be investigated first.

5.3.1 Metal-ferroelectric-metal-insulator-metal structure

The general electrostatic boundary conditions and electric fields for the MFMIM structure are identical to Eq. 5.9- 5.12, when substituting $P_S = (P_1 + P_2)/2$. This means that in the short-circuit condition and in the anti-parallel domain case $P_1 = -P_2$, the electric fields E_f and E_d vanish, which also results in $F_d = 0$. Therefore, there is no interaction between the ferroelectric and the dielectric layer anymore. The total energy of the system is thus given only by the contributions from the ferroelectric anisotropy and the domain wall energy as

$$F(P_1 = -P_2 = P_S) = t_f \left[\alpha + \left(\frac{2}{d} - \frac{1}{L} \right) \frac{4k}{w_{dw}} \right] P_S^2 + t_f \beta P_S^4. \tag{5.19}$$

Since all electric fields are zero, Eq. 5.19 is completely independent of ε_d, ε_b and t_d, which shows that the only way to prevent anti-parallel domain formation in the MFMIM structure, is through the domain wall energy term. Only the case of exactly two anti-parallel domains is of interest here, since this corresponds to the case of lowest domain wall energy. A larger number of domains (e.g. two up and two down domains) would not change the electrostatics, but would increase the domain wall energy, thus the formation of two domains is most favorable. Therefore, the total width of the two domains must be identical to the lateral capacitor size, i.e. $d = L$ (see Fig. 5.1). When setting $d = L$ in Eq. 5.19 and calculating the second derivative with respect to P_S, the NC stabilization condition for anti-parallel domains in the MFMIM structure is given by

$$\alpha + \frac{4k}{w_{dw}L} > 0. \tag{5.20}$$

The only parameter in Eq. 5.20 which is not a material constant is L, which means that a critical *lateral* dimension L_{crit} can be defined, below which anti-parallel domain formation will be suppressed:

$$L < L_{crit} = -\frac{4k}{\alpha w_{dw}}. \tag{5.21}$$

This means that to enable stabilized intrinsic NC in an MFMIM structure, all lateral dimensions would have to be extremely small to prevent domain formation. For typical perovskite ferroelectrics like $BaTiO_3$ or $PbTiO_3$, L_{crit} was estimated to be in the range of 1 nm, which is too small for practical devices [135]. However, recently, it was argued that a nanoscale MFMIM capacitor structure could exhibit extrinsic NC due to the reversible motion of a single domain wall in a two-domain ferroelectric [229]. Although, even in this case, an extremely small capacitor size is necessary [229]. Furthermore, leakage currents were neglected in this study, which have been shown to destabilize NC in the DC limit in any MFMIM structure [230]. The stabilizing nature of the domain wall energy when reducing the lateral dimension L can be seen from Fig. 5.3, where F is plotted as a function of P_1 and P_2 while the single-domain NC stabilization condition of Eq. 5.17 is fulfilled.

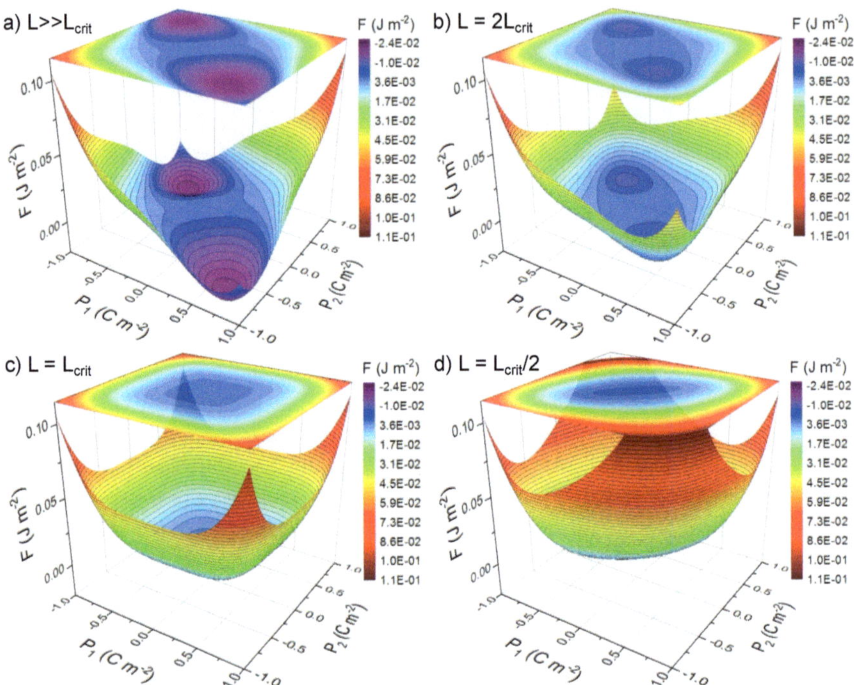

Figure 5.3: Total free energy density per area of a MFMIM structure with different lateral dimensions a) $L \gg L_{crit}$, b) $L = 2L_{crit}$, c) $L = L_{crit}$ and d) $L = L_{crit}/2$ as a function of the domain polarizations P_1 and P_2. Adapted from Ref. [135] with permission from The Royal Society of Chemistry.

In Fig. 5.3a), L is much larger than L_{crit}, which means that the domain wall energy contribution to the total energy is negligible. In this case, the degenerate energy minima corresponding to anti-parallel domains are identical to those in Fig. 5.1. However, the homogeneous energy minima have vanished due to the depolarizing effect of the dielectric layer. Nevertheless, since the $P_{1,2} = 0$ point is a saddle-point and not a global energy minimum, the intrinsic NC state is not stabilized. When the lateral dimensions are reduced to $L = 2L_{crit}$ as shown in Fig. 5.3b), the domain wall energy lifts the remaining energy minima corresponding to anti-parallel domain formation to higher energies. However, the $P_{1,2} = 0$ point is still unstable. Decreasing L further towards L_{crit} and below (Fig. 5.3c,d)) then increases the domain wall energy enough to make the anti-parallel domain states unstable. In this case, only the $P_{1,2} = 0$ point is the global energy minimum, which means that the intrinsic NC state is stabilized for $L \leq L_{crit}$. It can also be seen that the curvature of the energy landscape in the $P_1 = P_2$ direction is unchanged by L, which means that the voltage amplification due to intrinsic NC is not affected by the device size once it is stabilized.

To compare the domain wall energies of different ferroelectrics, Table 5.1 shows material parameters from literature and the calculated L_{crit}. The parameters for BaTiO$_3$ were taken from Kinase and Takahasi [231], Cao et al. [232] and Lu et al. [233]. Values for PbTiO$_3$ were taken from Behera et al. [234], Morozovska et al. [235] and Haun et al. [236]. The domain wall coupling constants k were obtained using the formula $k = F_{single,DW} w_{dw} / (4 P_S^2)$ where P_S is the spontaneous polarization of the anti-parallel domains in the ferroelectric and $F_{single,DW}$ is the domain wall energy density per domain wall area. Values for ferroelectric HfO$_2$ were estimated based on S. Clima et al. [98]. Domain wall energy constants for HfO$_2$ have not been reported so far. From Table S1 it can be seen that L_{crit} is in the order of 1 nm for perovskite ferroelectrics, which makes the use of an MFMIM structure impractical for NC devices as discussed in the main text. The domain coupling term $k/w_{dw} = 4.65 \times 10^{-2}$ m^2 F^{-1}, which was used for the simulation of the MFIM structure, is slightly smaller compared e.g. to PbTiO$_3$ with $k/w_{dw} = 5.83 \times 10^{-2}$ m^2 F^{-1}, but larger than BaTiO$_3$ with $k/w_{dw} = 7.13 \times 10^{-3}$ m^2 F^{-1}.

Table 5.1: Material parameters for typical ferroelectrics from literature and calculated critical lateral dimension L_{crit}.

Parameter	BaTiO$_3$	PbTiO$_3$	HfO$_2$	Units
$\alpha(T = 300\ \mathrm{K})$	-2.89×10^7	-1.72×10^8	-1×10^9	m F^{-1}
ε_b	45	67	25	1
k	2.85×10^{-12}	4.55×10^{-11}	?	m^3 F^{-1}
w_{dw}	4×10^{-10}	7.81×10^{-10}	5×10^{-10}	m
P_S	0.2216	0.867	0.5	C m^{-2}
$F_{single,DW}$	0.0014	0.175	?	J m^{-2}
L_{crit}	9.86×10^{-10}	1.36×10^{-9}	?	m

Based on these estimated values for L_{crit} and the previous discussion, it seems that the MFMIM structure is not suited to stabilize intrinsic NC. Furthermore, effects of leakage currents, which have been neglected here, might also complicate the use of extrinsic NC in such structures [230].

5.3.2 Metal-ferroelectric-insulator-metal structure

In contrast to the MFMIM case, where the electrostatics can be solved in only one dimension, for the MFIM case, 2D-electrostatics have to be solved due to the discontinuous boundary conditions at the ferroelectric/dielectric interface for anti-parallel stripe-domains. This leads to electric fields in the in-plane x-direction. An MFIM capacitor extending infinitely in the x-direction is considered, where anti-parallel domains with width d repeat periodically. To solve the electrostatics of the MFIM structure in the absence of magnetic fields, the starting point is Faraday's law of induction:

$$\nabla \times E = 0. \tag{5.22}$$

Eq. 5.22 is always valid for a scalar electrostatic potential φ, which is related to the electric field E by

$$E = -\nabla \varphi. \tag{5.23}$$

Next, Gauss's law is given by

$$\nabla \cdot D = \rho_{free}, \tag{5.24}$$

where D is the electric displacement field and ρ_{free} is the density of free charges per volume. Assuming a ferroelectric with relative permittivies $\varepsilon_{b,x}$ and $\varepsilon_{b,z}$ in the in-plane and out-of-plane directions, respectively (see Fig. 5.1), with spontaneous the polarization P_z only in z-direction the displacement field is given by

$$D = \varepsilon_0 \begin{pmatrix} \varepsilon_{b,x} & 0 \\ 0 & \varepsilon_{b,z} \end{pmatrix} E + \begin{pmatrix} 0 \\ P_z \end{pmatrix} = \begin{pmatrix} -\varepsilon_0 \varepsilon_{b,x} \frac{\partial \varphi}{\partial x} \\ -\varepsilon_0 \varepsilon_{b,z} \frac{\partial \varphi}{\partial z} + P_z \end{pmatrix}, \tag{5.25}$$

relating φ and P_z to D. If it is now assumed that there are no free charges in the structure ($\rho_{free} = 0$) and that $\varepsilon_{b,x}$ and $\varepsilon_{b,z}$ are functions of position, Eq. 5.25 can be inserted into Eq. 5.24 to obtain

$$\nabla \cdot D = \frac{\partial \varepsilon_{b,x}}{\partial x} \frac{\partial \varphi}{\partial x} + \frac{\partial \varepsilon_{b,z}}{\partial z} \frac{\partial \varphi}{\partial z} + \varepsilon_{b,x} \frac{\partial^2 \varphi}{\partial x^2} + \varepsilon_{b,z} \frac{\partial^2 \varphi}{\partial z^2} - \frac{1}{\varepsilon_0} \frac{\partial P_z}{\partial z} = 0. \tag{5.26}$$

Eq. 5.26 can now be solved for φ if $P_z(z, x)$, $\varepsilon_{b,x}(z, x)$ and $\varepsilon_{b,z}(z, x)$ as well as appropriate electrostatic boundary conditions are given. To numerically solve Eq. 5.26 a finite difference approach is applied, which approximates the derivatives of a function $f(x)$ as

$$\begin{aligned} \frac{\partial f}{\partial x} &\approx \frac{f_{x+1} - f_x}{\Delta x}, \\ \frac{\partial^2 f}{\partial x^2} &\approx \frac{f_{x+1} - 2f_x + f_{x-1}}{\Delta x^2}, \end{aligned} \tag{5.27}$$

where f_{x+1} and f_{x-1} are discretized values of f adjacent to the grid value f_x, where the derivative is taken and Δx is the numerical grid spacing in x-direction. Analogously to Eq. 5.27, the derivatives in z-direction are obtained by substituting x with z. Using this method to discretize Eq. 5.26, a formula for the electrostatic potential φ at the grid point (z, x) can be written as

$$\varphi_{zx} = \frac{(\varepsilon_x \varphi_{x-1} + \varepsilon_{x+1} \varphi_{x+1})\, \Delta z^2 + (\varepsilon_z \varphi_{z-1} + \varepsilon_{z+1} \varphi_{z+1})\, \Delta x^2 + (P_z - P_{z+1})\, \Delta z \Delta x^2 / \varepsilon_0}{(\varepsilon_x + \varepsilon_{x+1})\, \Delta z^2 + (\varepsilon_z + \varepsilon_{z+1})\, \Delta x^2}, \tag{5.28}$$

where Δz is the grid spacing in z-direction. Indices in Eq. 5.28 indicate the grid position relative to (z, x). In addition to Eq. 5.28 electrostatic boundary conditions are needed to calculate φ. By defining a numerical grid of dimensions $Z \times X$, where $z = 1, 2, ..., Z$ and $x = 1, 2, ..., X$, $z = 1$ indicates the top and $z = Z$ the bottom electrode of the capacitor structure. For the short-circuit condition of both electrodes one obtains

$$\varphi(z = 1, x) = \varphi(z = Z, x) = 0. \tag{5.29}$$

Furthermore, to simulate an infinite capacitor structure in x-direction, periodic boundary conditions at the remaining edges of the grid are applied:

$$\begin{aligned} \varphi(z, x = 0) &= \varphi(z, x = X), \\ \varphi(z, x = 1) &= \varphi(z, x = X + 1). \end{aligned} \tag{5.30}$$

To solve Eq. 5.28, an iterative algorithm was used where the electrostatic potential of the current iteration $\varphi^{(i)}$ is calculated from the value of the last iteration $\varphi^{(i-1)}$, as schematically shown in Fig. 5.4. At each iteration step (i), the remaining error at each grid point is calculated as

$$\epsilon^{(i)} = \left| \frac{\varphi^{(i-1)} - \varphi^{(i)}}{\varphi^{(i-1)}} \right|. \tag{5.31}$$

This process is repeated until the maximum error for φ is below 0.01 % or

$$\max(\epsilon^{(i)}) < 10^{-4}. \tag{5.32}$$

From φ, E and D are then easily obtained from Eq. 5.23 and Eq. 5.25, respectively. The free energy density per unit volume at each grid point (z, x) is then calculated as

$$F_{f/d}(z,x) = \alpha P_z^2 + \beta P_z^4 + k \left(\frac{P_z(z,x) - P_z(z,x+1)}{\Delta x} \right)^2 + \frac{1}{2} ED, \qquad (5.33)$$

where P_z is only the spontaneous polarization, which is zero in the dielectric region. To calculate the average free energy per domain period of an infinite capacitor in x- and y-direction per unit area one can use the equation

$$F = \frac{1}{d} \sum_{x=1}^{X} \sum_{z=1}^{Z} F_{f/d}(z,x)\Delta x \Delta z, \qquad (5.34)$$

where $d = X\Delta x$ is the domain period as defined in Fig. 5.1b).

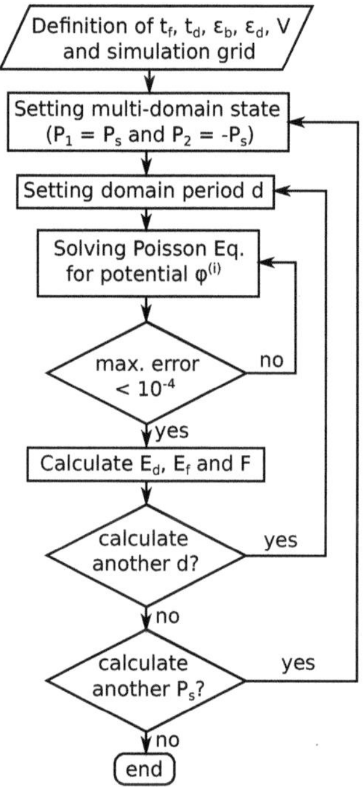

Figure 5.4: Flowchart for the multidomain Ginzburg-Landau and electrostatics simulations of the MFIM structure in Fig. 5.5. Adapted from Ref. [228]. © 2018 IEEE.

The simulation parameters used in the following were $\alpha = -4 \times 10^7$ m F^{-1}, $\beta = 4.2 \times 10^7$ m^5 C^{-2} F^{-1}, $k/w = 4.65 \times 10^{-2}$ m^2 F^{-1}, $X = 100$, $\Delta x = d/X$, $\Delta z = 5 \times 10^{-11}$ m and $\varepsilon_{b,x} = \varepsilon_{b,x} = \varepsilon_b = 30$. For the dielectric, $\varepsilon_d = 400$ was used, which is close to the relative permittivity of SrTiO$_3$ [237]. It should be noted that Z was changed depending on the total thickness $t_D + t_f$ to keep Δz constant for all simulations. This ensured a clearly defined grid position of the ferroelectric/dielectric interface. Furthermore, for all simulations it was checked that Δx and Δz are sufficiently small such that no change of the electrostatics was observed when reducing them further.

To investigate the stability of the intrinsic NC state in the MFIM stack against domain formation, the total free energy density per domain period is calculated for various domain periods d and anti-parallel spontaneous polarization values $P_1 = -P_2$. In Fig. 5.5a), the resulting free energy landscape is shown for a 10 nm thin ferroelectric film. Two degenerate energy minima can be observed at non-zero spontaneous polarization for certain values of d. In this case, the anti-parallel multi-domain state is most favorable, since it has the lowest energy at the equilibrium domain width $d = d_{eq}$, balanced by the competing depolarization and domain wall energies. For larger domain widths ($d > d_{eq}$), the depolarization energy increases, eventually leading to a vanishing polarization. For smaller domain widths, on the other hand, a strong increase of the domain wall energy leads to a suppression of domain formation. However, since an infinite film in x-direction is assumed here, the ferroelectric is only thermodynamically stable at the equilibrium domain width d_{eq}.

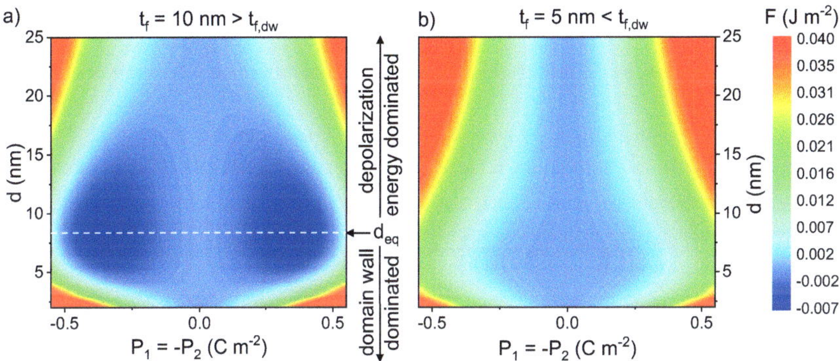

Figure 5.5: Total free energy of an infinite MFIM structure with anti-parallel domains as a function of the domain period d and the polarization a) for $t_f > t_{f,dw}$ and b) $t_f < t_{f,dw}$. Adapted from Ref. [135] with permission from The Royal Society of Chemistry.

When reducing the ferroelectric thickness to 5 nm while keeping all other parameters constant, the free energy landscape changes as shown in Fig. 5.5b). In this case, the energy minima corresponding to anti-parallel domain formation completely vanish, leading to global minimum at $P_1 = P_2 = 0$ independent of the domain width d. This means, that the ferroelectric is completely depolarized and the intrinsic NC state is stable. It should be noted that for both simulations in Fig. 5.5, the condition $t_f < t_{f,max}$ was fulfilled, which means that the single-domain critical thickness for intrinsic NC stabilization is typically much larger than the critical thickness for multi-domain stabilization, which will be called $t_{f,dw}$. To obtain an analytic formula for $t_{f,dw}$, different material parameters α, k/w_{dw} and $\varepsilon_d + \varepsilon_b$ were used for the numerical simulations as shown in Fig. 5.6.

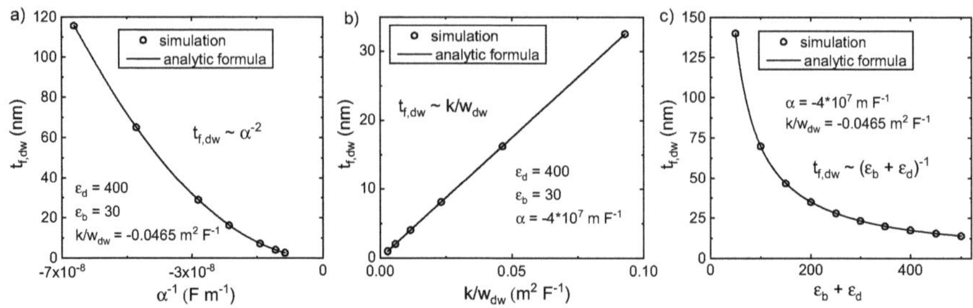

Figure 5.6: Comparison of simulation results and the analytic formula for $t_{f,dw}$ as a function of a) α, b) k/w_{dw} and c) $\varepsilon_b + \varepsilon_d$. Adapted from Ref. [228]. © 2018 IEEE.

Then, the different dependencies of $t_{f,dw}$ were fitted to an analytical model which resulted in the following condition for prevention of anti-parallel domain formation:

$$t_f < t_{f,dw} \approx \frac{2k}{\alpha^2 \varepsilon_0 (\varepsilon_b + \varepsilon_d) w_{dw}} \tag{5.35}$$

This expression is similar to the critical thickness for sinusoidal domain formation obtained by Cano and Jiménez [238]. Furthermore, it should be noted that Eq. 5.35 is independent of t_d, as long as t_d is large enough compared to the equilibrium domain width $d_{eq}/2$ [7, 134]. This can also be seen from the simulation results in Fig. 5.7, which shows the critical thickness for domain formation $t_{f,dw}$ as a function of t_d.

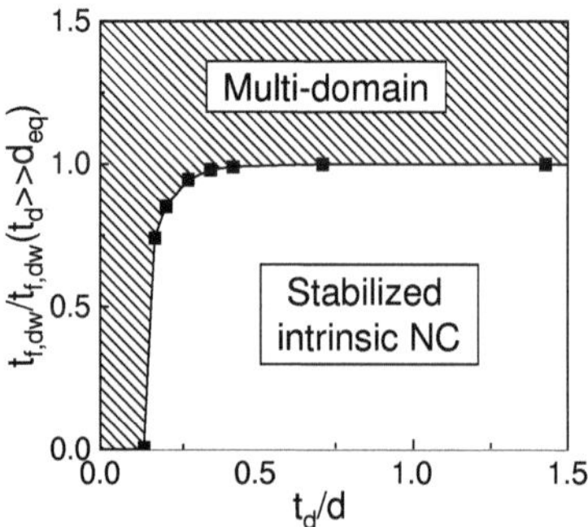

Figure 5.7: Stability of the multidomain state as a function of t_d normalized to the equilibrium domain period d. $t_{f,dw}$ is the critical thickness for the stabilization of the NC state when domains can form. Adapted from Ref. [228]. © 2018 IEEE.

As can be seen, the critical thickness $t_{f,dw}$ quickly drops to zero when t_d becomes smaller than $d_{eq}/2$. The reason is that below this thickness, the electric field lines inside the dielectric do not close completely, and part of the bound ferroelectric polarization is compensated by free charges in the metal electrode. On the other hand, if the dielectric is much thicker than $d_{eq}/2$, the electric field lines would close mainly in a region close to the ferroelectric/dielectric interface, making the rest of the dielectric field-free [134]. Therefore, in this region where t_d is thicker than $d_{eq}/2$, $t_{f,dw}$ is independent of t_d as described by Eq. 5.35.

Now using the ferroelectric parameters for BaTiO$_3$ and PbTiO$_3$ from Table 5.1, the single-domain and multi-domain critical ferroelectric thicknesses as well as the critical lateral dimension L_{crit} are calculated as a function of the permittivity of the dielectric layer (see Fig. 5.8). In general, it can be seen that the critical thickness for domain formation $t_{f,dw}$ is much lower than the single-domain critical thickness $t_{f,max}$. This is why the consideration of domain formation is crucial for the stabilization even of intrinsic NC [238]. Furthermore, for lower permittivity dielectrics, $t_{f,dw}$ does only slightly change due to the non-zero background permittivity of the ferroelectric. The critical lateral dimensions are in the range of 1 nm, as discussed before.

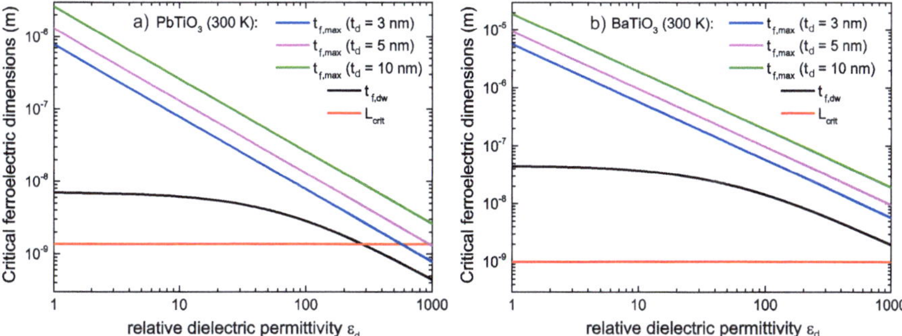

Figure 5.8: Critical ferroelectric dimensions $t_{f,dw}$, $t_{f,max}$ and L_{crit} in a MFIM capacitor as a function of ε_d for different dielectric thicknesses t_d for a) PbTiO$_3$ and b) BaTiO$_3$ material parameters taken from Table 5.1. Adapted from Ref. [135] with permission from The Royal Society of Chemistry.

Comparing BaTiO$_3$ and PbTiO$_3$, the critical thickness $t_{f,dw}$ is much larger for the former, since α^2 is much smaller at 300 K. This means that some materials might be more favorable for NC applications based on their intrinsic material parameters, like Landau coefficients and domain wall energy constants.

5.4 Summary

When investigating the stabilization of intrinsic NC in a ferroelectric, the possibility of domain formation must be considered. If only single-domain models are used, qualitatively different results can occur. In particular, intrinsic NC stabilization in MFMIM structures would only be feasible for extremely small lateral device dimensions in the range of 1 nm, due to the formation of anti-parallel domains, see Fig. 5.9a). Therefore, the MFIM structure without internal metal electrode is mandatory for building hysteresis-free NC devices. In such MFIM structures, the critical thickness for intrinsic NC stabilization is much smaller when considering domain formation, thus limiting the maximum attainable voltage amplification of NC devices, see Fig. 5.9b). However, the critical thickness for NC stabilization as well as the maximum voltage amplification might be increased by reducing the lateral device dimensions towards the nanoscale, see Fig. 5.9b). Based on these insights, in the following Chapter, MFIM capacitors are experimentally investigated for stabilized NC effects using HfO_2 based ferroelectrics and short pulsed voltage measurements.

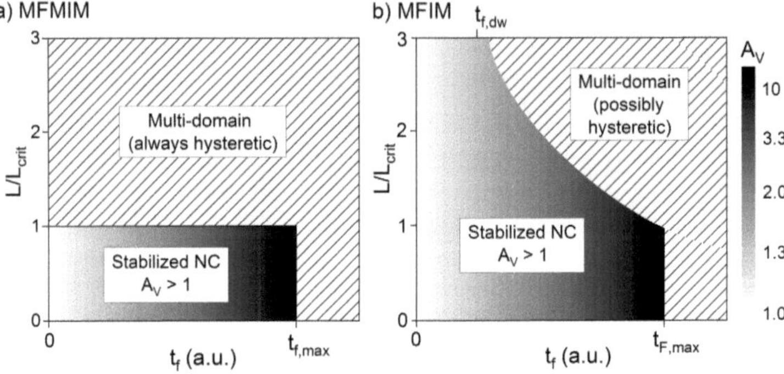

Figure 5.9: Voltage amplification A_V and intrinsic NC stabilization regime as a function of the ferroelectric thickness t_f and the lateral dimension L for a) the MFMIM and b) the MFIM structure. Adapted from Ref. [135] with permission from The Royal Society of Chemistry.

6 Stabilized negative capacitance under short pulsed voltage operation

Since the direct measurement of transient NC in ferroelectric capacitors with metal electrodes inevitably results in a polarization-electric field hysteresis as discussed in Section 4, other experimental approaches are necessary to obtain hysteresis-free (i.e. stabilized) NC behavior. In particular, inserting an insulating layer between the ferroelectric and one or both of the metal electrodes can prevent the screening of the bound polarization charge, thus suppressing hysteretic ferroelectric switching and potentially stabilizing NC, see the schematic in Fig. 6.1. While stabilized NC behavior has been previously reported in high-quality epitaxial perovskite heterostructures and superlattices [7, 6, 239, 240], similar observations in HfO_2 based ferroelectrics have remained elusive.

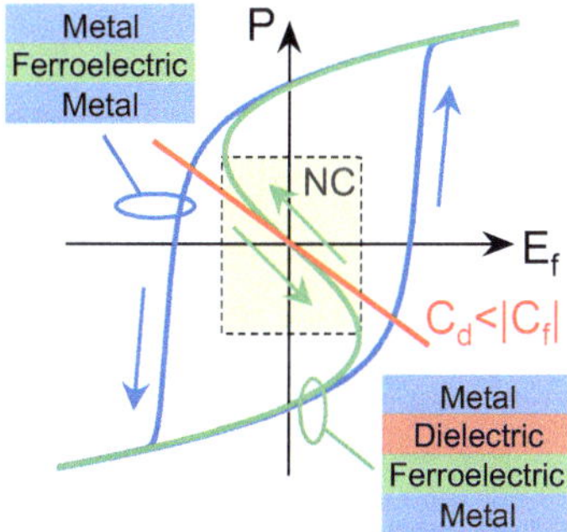

Figure 6.1: Polarization-electric field (P-E_f) dependence of a ferroelectric based on Landau theory. Without dielectric, hysteresis and only positive capacitance C_f is expected while with dielectric, no hysteresis and NC $C_f < 0$ can be achieved, if the dielectric capacitance C_d is matched in the NC region, i.e. $C_d < |C_f|$ (see red load line of the dielectric). Adapted from Ref. [241]. © 2018 IEEE.

In the first part of this chapter, the experimental demonstration of stabilized NC in heterostructure capacitors with HfO_2 based ferroelectrics and Al_2O_3 dielectrics under pulsed voltage operation is described, which was published in Ref. [241]. These results show indications of the typical 'S'-shaped P-E_f curve predicted by Landau theory. In the second part, Ta_2O_5 was applied as a dielectric to extend the accessible range of the measured 'S'-shaped curve, allowing the extraction of the double-well energy landscape of the ferroelectric layer, which was published in Ref. [128]. Finally, based on the previous experimental results, a way to overcome the fundamental limits of energy storage in capacitors is proposed, which could be applied to create a new type of solid-state electrostatic supercapacitor utilizing NC [242].

6.1 Ferroelectric/dielectric $Hf_{0.5}Zr_{0.5}O_2$/Al_2O_3 capacitors

To investigate the potentially hysteresis-free NC behavior in ferroelectrics based on HfO_2, ferroelectric/dielectric heterostructure capacitors with different layer thicknesses were fabricated. Due to the wide process window, low crystallization temperature and robust properties [9], ferroelectric HZO with two different thicknesses of 7.7 nm and 11.3 nm was used. For the dielectric layer, amorphous Al_2O_3 was preferred due to its large electronic band gap of 6.2 eV [243], relatively high permittivity and high breakdown field strength [244]. Additionally, the ALD of Al_2O_3 could be carried out in the same chamber without vacuum break after the ALD of HZO, reducing possible interface contaminations. The Al_2O_3 thickness was varied from zero to 4 nm. For the characterization of the ferroelectric properties of the HZO layers, samples without Al_2O_3 were fabricated. This section is based on the results published in Ref. [241].

6.1.1 Standard ferroelectric characterization

After crystallization of the HZO films into the ferroelectric phase and structuring of the capacitor top electrodes, GIXRD measurements were carried out. The measured GIXRD patterns for the two different HZO thicknesses are shown in Fig. 6.2 together with reference patterns for the orthorhombic, monoclinic and tetragonal phases in HfO_2.

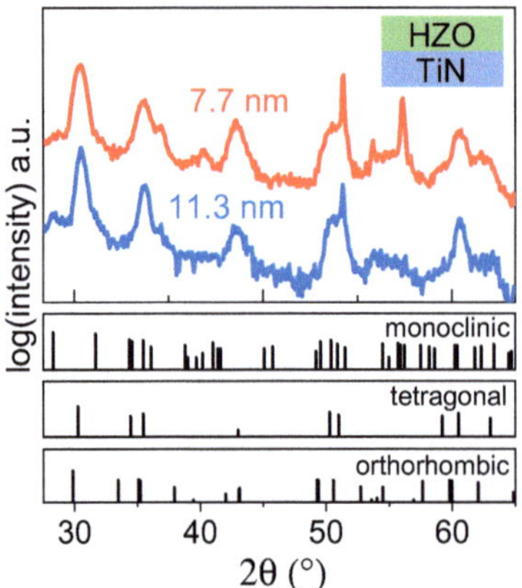

Figure 6.2: GIXRD patterns of ferroelectric HZO layers of different thickness compared to reference patterns for common phases in HfO_2. The HZO films are mostly orthorhombic. Adapted from Ref. [241]. © 2018 IEEE.

Both films show only small amounts of monoclinic phase fraction, while the fraction in the thicker film seems to be slightly larger based on the peak around 28°, which is in agreement with previous reports that the monoclinic phase fraction increases for larger film thickness in HZO thin films [245].

The measured patterns closely resemble the reference patterns of the ferroelectric orthorhombic and tetragonal phases. In a next step, standard electrical characterization of the TiN/HZO/TiN capacitors was carried out. Fig. 6.3 shows the P-E_f hysteresis loops measured at a frequency of 10 kHz after applying 10^5 rectangular preconditioning cycles at 100 kHz due to the wake-up effect [246, 247, 248]. The remanent polarization for the 7.7 nm and 11.3 nm thin HZO layers are 25 µC cm^{-2} and 27 µC cm^{-2}, respectively. These comparatively high values are in agreement with almost exclusively orthorhombic films as suggested by the GIXRD data in Fig. 6.2. The coercive fields for both thicknesses have roughly identical values of $\sim$1 MV cm^{-1} in agreement with literature [82, 245].

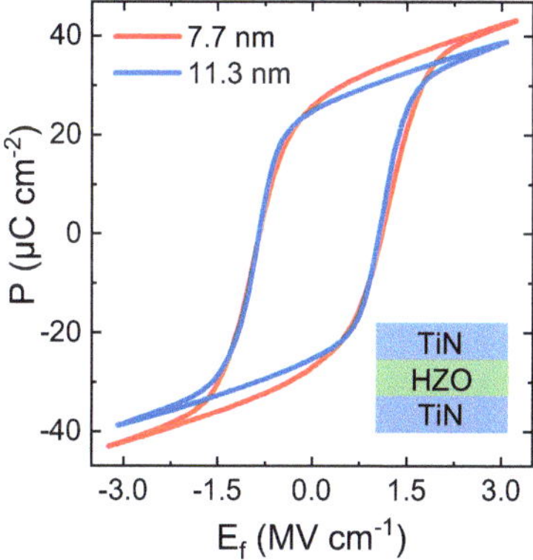

Figure 6.3: Experimental P-E_f hysteresis curves measured at 10 kHz for TiN/HZO/TiN capacitors with different HZO thicknesses after wakeup cycling (10^5 cycles at 100 kHz). Adapted from Ref. [241]. © 2018 IEEE.

Small-signal capacitance-voltage (C-V) measurements confirm excellent ferroelectric properties leading to symmetric 'butterfly' curves, which are shown in Fig. 6.4. These measurements were also done after 'wake-up' cycling with a small-signal frequency of 10 kHz and an AC-amplitude of 50 mV. Extracted relative background permittivities are roughly 39 and 37 for the 7.7 nm and 11.3 nm thin film, respectively, which are slightly larger than previously reported values from literature ($\sim$28-33) [97, 142].

Of further interest is the stability of the ferroelectric properties when applying a large number of switching cycles. Fig. 6.5 shows the change of the remanent polarization P_r in pristine samples after applying bipolar voltage pulses with a frequency of 100 kHz and amplitudes of 2.5 V and 3 V for the 7.7 nm and the 11.3 nm sample, respectively. The initial increase in P_r until roughly 10^5 cycles is related to the wake-up effect, which is stronger in the thinner HZO sample in agreement with literature results [249]. Beyond 10^5 cycles P_r stays relatively constant with only small indications of polarization fatigue in the 11.3 nm sample, where P_r slightly decreases for higher cycling numbers.

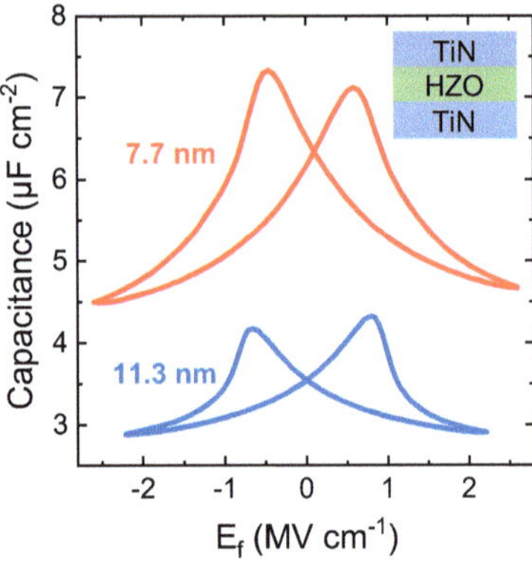

Figure 6.4: Capacitance-electric field hysteresis for different HZO thicknesses measured at the frequency $f =$ 10 kHz and 50 mV small-signal amplitude after wake-up cycling (10^5 cycles at 100 kHz). Adapted from Ref. [241]. © 2018 IEEE.

Overall, both the 7.7 nm and 11.3 nm thin HZO films exhibit excellent ferroelectric properties as indicated by a high orthorhombic phase fraction, a large P_r, symmetric butterfly C-V curves and stable cycling endurance. In a next step, HZO capacitors with intermediate Al_2O_3 layers of different thickness were analyzed.

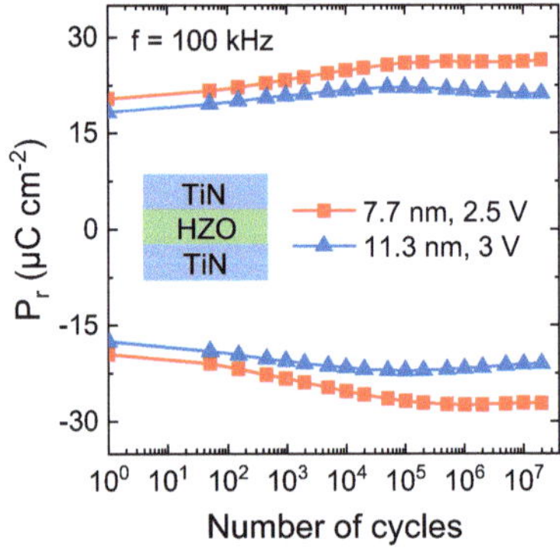

Figure 6.5: Evolution of remanent polarization P_r with electric field cycling ($f = 100$ kHz) for different HZO thicknesses. Adapted from Ref. [241]. © 2018 IEEE.

6.1.2 Basic $Hf_{0.5}Zr_{0.5}O_2/Al_2O_3$ capacitor characterization

The basic structure of one of the TiN/HZO/Al_2O_3/TiN heterostructure capacitors with 7.7 nm HZO and about 4 nm Al_2O_3 can be seen in the TEM image in Fig. 6.6a). Besides the confirmation of the layer thicknesses, the high-resolution TEM in Fig. 6.6b) shows the polycrystalline nature of the

HZO layer as well as the amorphous morphology of the Al_2O_3 layer. No significant intermixing between HZO and Al_2O_3 can be seen from the TEM results.

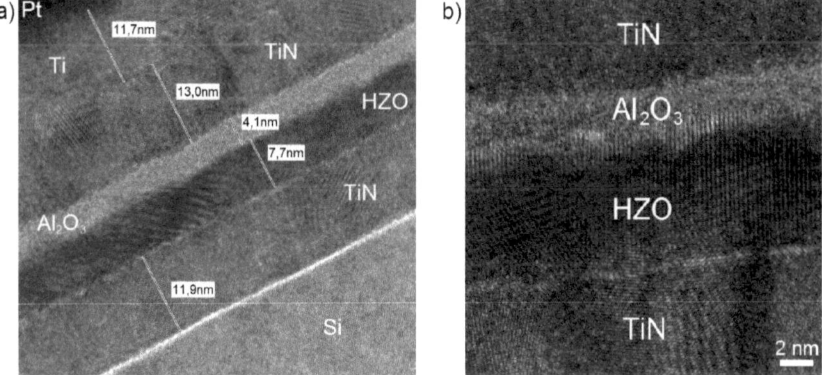

Figure 6.6: a) TEM cross-section and b) high-resolution TEM of an Al_2O_3/HZO capacitor structure with about 4 nm Al_2O_3 and 7.7 nm HZO. The polycrystalline structure of HZO and amorphous phase of the Al_2O_3 layer can be seen. Adapted from Ref. [241]. © 2018 IEEE.

To determine if any of the HZO layers are stabilized in the NC state, small-signal capacitance measurements were carried out at zero DC bias. The relative permittivity of the Al_2O_3 was determined as ~8 from the measurement of a separate TiN/Al_2O_3/TiN capacitor. Fig. 6.7 depicts the inverse of the small-signal capacitance for all samples as a function of the Al_2O_3 thickness. The green line indicates the expected inverse capacitance for an Al_2O_3 capacitor with a relative permittivity of 8. The red and blue lines correspond to linear fitting curves for the 7.7 nm and 11.3 nm HZO samples, respectively. If any of the HZO layers would be in a stabilized NC state, then the total inverse capacitance would have to be smaller than that of the Al_2O_3 layer alone. However, since all measurement points clearly are above the green line, none of the samples shows stabilized NC at zero DC bias. In fact, the samples behave exactly as expected from two positive capacitance layers in series as given by the equation

$$\frac{1}{C} = \frac{1}{C_f} + \frac{t_d}{\varepsilon_0 \varepsilon_d}, \tag{6.1}$$

where C_f is the capacitance of HZO and t_d and ε_d are the thickness and relative permittivity of Al_2O_3, respectively. When fitting both C_f and ε_d to the experimental data, excellent agreement with Eq. 6.1 can be seen.

The reason for the absence of stabilized NC at zero bias can be explained by trapped electrons at the HZO/Al_2O_3 interface [211, 250]. If the density of trapped electrons at the interface is comparable to the spontaneous polarization of the HZO, the depolarization field will become negligible, thus stabilizing the polarization in the direction pointing towards the interface. While it is known that

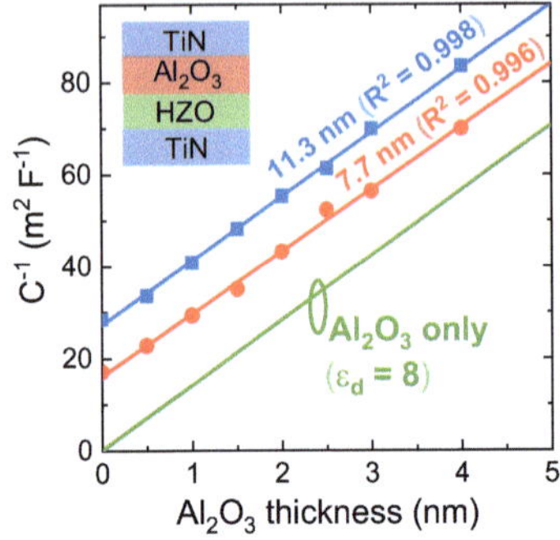

Figure 6.7: Inverse small-signal capacitance C (f = 10 kHz, 50 mV amplitude) for different HZO and Al_2O_3 thicknesses. Solid blue and red lines show linear fits and the green line corresponds to the theoretical capacitance of Al_2O_3 without HZO. Adapted from Ref. [241]. © 2018 IEEE.

negative fixed charges can form at the interface between Al_2O_3 and other dielectric layers [251], this charge density is roughly an order of magnitude lower than the spontaneous polarization of HZO thin films. Therefore, it is more likely that additionally electrons are trapped at deep trap states at the HZO/Al_2O_3 interface [252].

6.1.3 Pulsed charge-voltage measurement results

As was proposed by Kim et al. [211], to access the stabilized NC region, applying short voltage pulses is necessary such that the spontaneous polarization during switching is not screened by free charges, which would lead to hysteresis [253]. The experimental setup for such an experiment is shown schematically in Fig. 6.8.

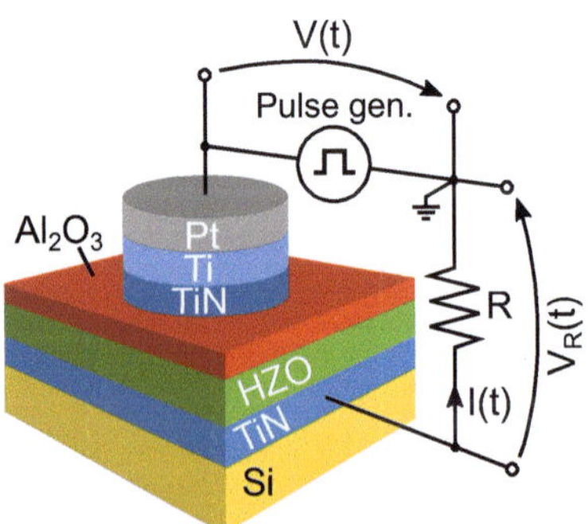

Figure 6.8: Schematic sample structure and experimental setup for pulsed charge-voltage measurements. Short voltage pulses are applied to the top electrode while the current is measured through the 50 Ω input resistance of an oscilloscope. The pulse voltage is measured at the same time. Adapted from Ref. [241]. © 2018 IEEE.

A pulse generator is connected to the top electrode of the heterostructure capacitor to apply a voltage $V(t)$, which is simultaneously measured with an oscilloscope. The current $I(t)$ is measured via another oscilloscope channel with 50 Ω input resistance as the voltage $V_R(t)$ through the bottom electrode. The applied voltage pulses of constant pulse width ($\sim$500 ns) and rise/fall times ($\sim$150 ns) and increasing pulse amplitude are shown in Fig. 6.9a) for the sample with 4 nm Al_2O_3 and 11.3 nm HZO.

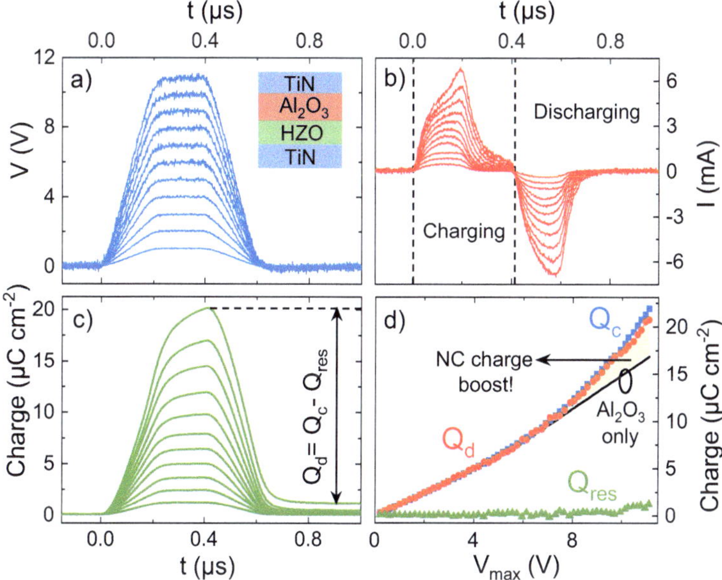

Figure 6.9: a) Applied voltage pulses with increasing amplitude. b) Measured current during charging and discharging of the 11.3 nm HZO/4 nm Al_2O_3 capacitor. c) Charge on the capacitor as a function of time t integrated from the current in b). d) Maximum, released and residual charges (Q_c, Q_d and Q_{res}) as a function of the maximum applied voltage. Black line shows expected charge for 4 nm Al_2O_3 layer without HZO. Adapted from Ref. [241]. © 2018 IEEE.

The measured current during charging and discharging of the capacitor with an area of roughly $\sim$7000 µm^2 is shown in Fig. 6.9b) for different pulse amplitudes. The charge on the capacitor as a function of time is calculated by integrating the current, which is plotted in Fig. 6.9c. As was proposed by Kim et al. [211], the three important charges for each voltage pulse amplitude are the maximum charge Q_c, the charge flowing during discharging Q_d and the residual charge Q_{res}, which is equal to the difference between the two, see Fig. 6.9c). In Fig. 6.9d), all three characteristic charges are shown as a function of the voltage pulse amplitude V_{max} from the data in Fig. 6.9a).

While Q_c also includes the flowing charge due to leakage currents, Q_d should only correspond to the capacitive behavior of the sample. Therefore, the overall differential capacitance of the capacitor is given by $C = dQ_d/dV_{max}$. As can be seen from Fig. 6.9d), the capacitance is roughly constant from 0 to 5 V and then strongly increases for higher voltages. At about 6 V, the total differential capacitance is actually larger than the differential capacitance of the Al_2O_3 layer alone, which is indicated by the black line in Fig. 6.9d). According to Eq. 6.1, this means that the capacitance of the HZO layer is negative in this region. Interestingly, this was only observed for positive applied pulses, but not for negative ones. In the latter case, only a constant linear slope of the Q_d-V_{max} curve was observed. By using the following expression, the field inside the ferroelectric can be calculated as

$$E_f = \frac{1}{t_f} \left(\max \left[V(t) - I(t)R \right] - \frac{Q_d}{C_D} \right), \tag{6.2}$$

where C_D is the capacitance of the Al_2O_3 layer in this case. With $\varepsilon_d = 8$ and a thickness of 4 nm, $C_D/A = 1.77 \ \mu F \ cm^{-2}$. Assuming a fixed charge σ_{IF} at the interface between HZO and Al_2O_3, the ferroelectric polarization can be calculated as

$$P = \frac{Q_d}{A} - \varepsilon_0 E_f + \sigma_{IF} \approx \frac{Q_d}{A} + \sigma_{IF}. \tag{6.3}$$

When assuming $\sigma_{IF} = -18 \ \mu C \ cm^{-2}$, the calculated P-E_f curve has the shape shown in Fig. 6.10. This data indicates a region of NC marked by the dashed box and yellow background.

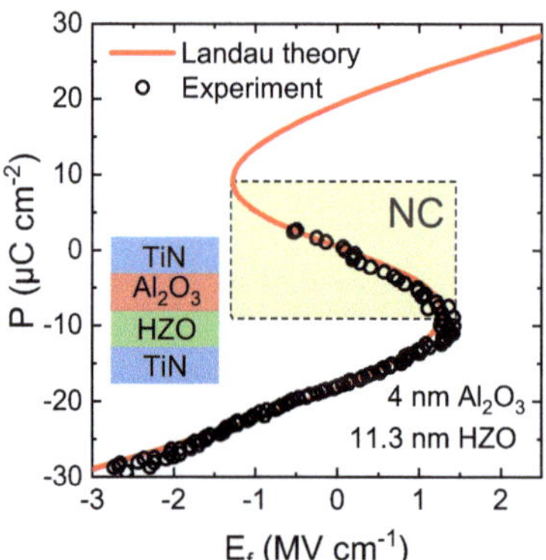

Figure 6.10: Ferroelectric P-E_f curve of 11.3 nm HZO with 4 nm Al_2O_3 extracted from the pulsed charge-voltage measurements in Fig. 6.9. Initially, the ferroelectric is in the negative remanent state with $P_r = -18 \ \mu C \ cm^{-2}$. For negative applied pulses, only a linear dielectric response in observed, since the HZO is already switched to negative polarization. However, for positive applied pulses, the ferroelectric enters the NC region corresponding to the negative slope of the 'S'-shaped Landau P-E_f curve. Adapted from Ref. [241]. © 2018 IEEE.

In the same graph, the solid red line presents a homogeneous Landau model fitted to the experimental data. Here, $\alpha = -8.8 \cdot 10^8$ m F^{-1}, $\beta = 1.3 \cdot 10^{10}$ m^5 C^{-2} F^{-1} and a background relative permittivity $\varepsilon_b = 25$ was used. Note that, for the highest polarization values, E_f even becomes negative, which means that the voltage drop across the Al$_2$O$_3$ will be larger than the externally applied voltage (absolute voltage amplification). This also results in a very high electric field inside the Al$_2$O$_3$ layer, which ultimately leads to hard dielectric breakdown for the highest applied voltages. This is the reason, why the 'S'-shaped curve cannot be completely measured in this sample.

6.1.4 Investigation of hysteresis

Another important aspect of ferroelectric NC for potential applications is the absence of polarization hysteresis. Therefore, additional experiments were carried out, where the amplitude of the applied voltage pulses was first successively increased until a maximum value and then decreased in the same manner, e.g. 0.2 V, 0.4 V, ... 9.8 V, 10 V, 9.8 V, ..., 0.4 V, 0.2 V. In this way it was investigated if the resulting Q_d-V_{max} curve exhibits a hysteresis. As can be seen in Fig. 6.11, no hysteresis was observed for the samples with 4 nm Al$_2$O$_3$ for both 7.7 nm and 11.3 nm thin HZO layers.

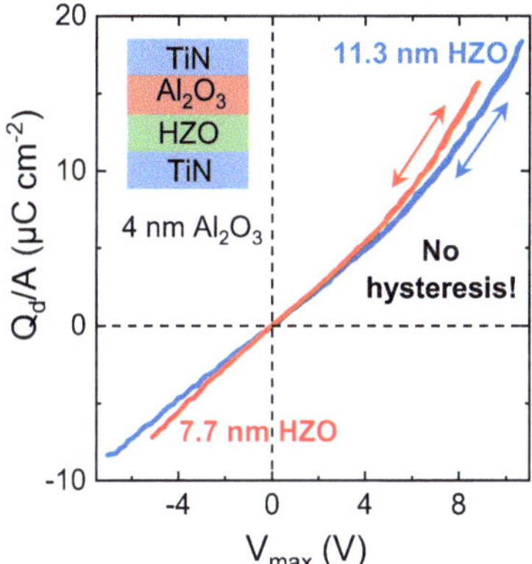

Figure 6.11: Pulsed charge-voltage hysteresis measurement for capacitors with 4 nm Al$_2$O$_3$. No hysteresis is observed for ascending and descending voltage pulse trains with 500 ns pulse width. Adapted from Ref. [241]. © 2018 IEEE.

These results demonstrate, that no hysteretic polarization switching is happening, because no negative reset pulses were applied between the positive measurement pulses. Indeed, when plotting the ferroelectric P-E_f curve for these measurements, the 'S'-shaped curve is traced for both increasing and decreasing pulse amplitudes as shown in Fig. 6.12 for the sample with 4 nm Al$_2$O$_3$ and 7.7 nm HZO thickness. For the Landau fitting curve, $\alpha = -1.1 \cdot 10^9$ m F^{-1}, $\beta = 2.5 \cdot 10^{10}$ m^5 C^{-2} F^{-1} and a background relative permittivity $\varepsilon_b = 25$ was used. This shows that the NC state at higher applied voltages for short voltage pulses is stabilized by the dielectric capacitance. The necessity of short measurement pulses does not necessarily mean that the measured NC effect is inherently transient by itself, but rather that the measurement time must be kept small due to practical external limitations of charge trapping and dielectric breakdown [211, 254]. However, further investigations are necessary to confirm that hysteresis-free NC operation is also possible for much longer applied voltage signals in optimized ferroelectric/dielectric structures.

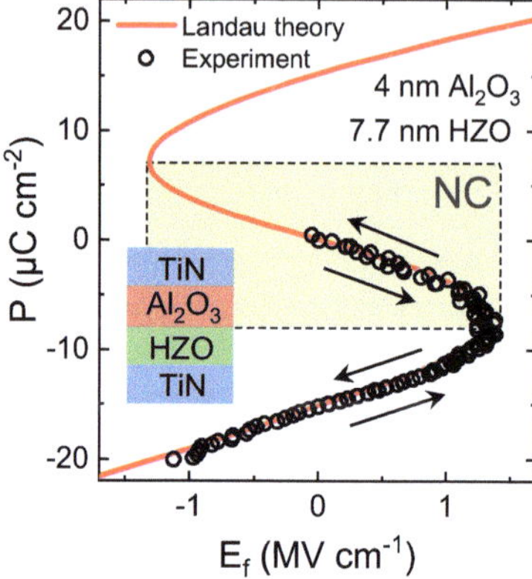

Figure 6.12: Extracted P-E_f curve for 7.7 nm HZO with 4 nm Al$_2$O$_3$ from Fig. 6.11. No hysteresis is observed in the 'S'-shaped curve in accordance with Landau theory. Adapted from Ref. [241]. © 2018 IEEE.

The reason for the absence of hysteresis seems to be the inhibition of charge compensation during switching related to the relatively thick dielectric layer (low leakage currents) and the short pulse time (less time for charges to tunnel to the interface). To investigate the effect of the layer thickness, a similar hysteresis measurement as before was carried out on the sample with only 1 nm Al$_2$O$_3$ and 7.7 nm HZO. The resulting charge-voltage curves are shown in Fig. 6.13.

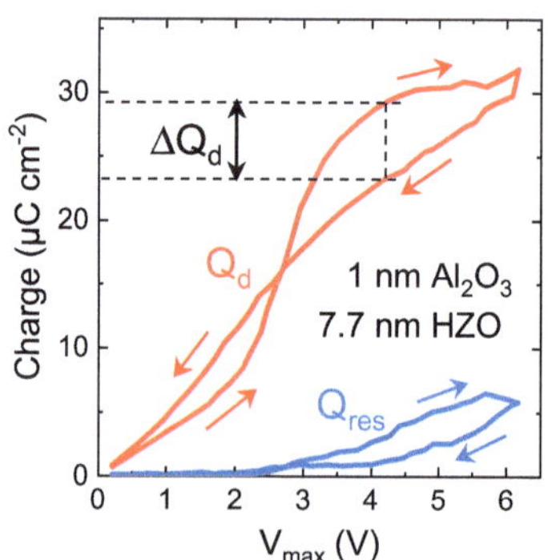

Figure 6.13: Pulsed charge-voltage measurement for a capacitor with only 1 nm Al$_2$O$_3$. Hysteresis is observed due to capacitance mismatch and/or charge trapping. ΔQ_d is defined as the maximum Q_d hysteresis. Adapted from Ref. [241]. © 2018 IEEE.

In contrast to the hysteresis-free characteristics with 4 nm thin Al_2O_3, the sample with only 1 nm Al_2O_3 shows a strong hysteresis in both Q_d and Q_{res} indicating at least partial ferroelectric switching. In addition, Q_{res} in general is much larger compared to the samples with thicker Al_2O_3, since free charges can easily tunnel through the layer for lower thicknesses. To better quantify this effect, the maximum hysteresis in Q_d was defined as the maximum hysteresis window ΔQ_d. Fig. 6.14 shows this hysteresis window as a function of both the HZO and Al_2O_3 thickness.

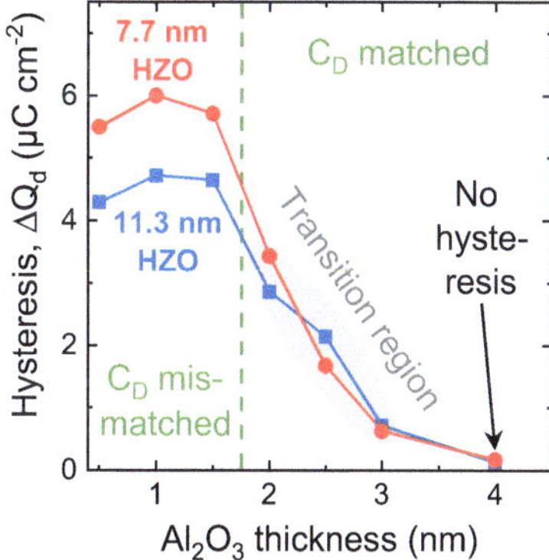

Figure 6.14: Maximum hysteresis as a function of layer thicknesses. For Al_2O_3 layers thicker than 3 nm, no hysteresis is observed in the pulsed charge-voltage measurements with pulse widths of 500 ns. The transition region is due to partial domain mismatch and/or charge trapping effects. Adapted from Ref. [241]. © 2018 IEEE.

The general trend of increasing hysteresis with decreasing Al_2O_3 thickness is consistent with both capacitance matching from Landau theory and charge injection to the HZO/Al_2O_3 interface due to leakage through the Al_2O_3 layer. Based on the extracted Landau parameters for HZO from Fig. 6.10 and Fig. 6.12, the ideal capacitance matching condition for Al_2O_3 would be around 1.8 nm thickness. From homogeneous Landau theory (neglecting leakage), it would be therefore expected that for Al_2O_3 thicknesses larger than 1.8 nm, no hysteresis is observed. However, Fig. 6.14 still shows considerable hysteresis for thicknesses between 2 and 3 nm, which only vanishes for the 4 nm samples. There are two possible explanations for this transition region: (1) Due to the polycrystalline nature of the HZO, different grains will have their ideal matching conditions for different Al_2O_3 thicknesses. Since the measured 'S'-curves in Fig. 6.10 and Fig. 6.12 only represent the average of all the grains, some of them will still be mismatched beyond 1.8 nm thickness. (2) The injection and trapping of charges at the HZO/Al_2O_3 interface can promote ferroelectric switching [211], even when all grains would be stabilized in the hysteresis-free NC region according to Landau theory. Likely, a combination of both effects plays a role in this transition region.

Another important aspect of ferroelectric NC is its potential speed limit. While there have been discussions that the fundamental speed limit of NC should not be much different to the response speed of currently used high-k dielectrics [49], experimental data is scarce [196]. Therefore, using even shorter voltage pulses was of interest to investigate whether hysteresis-free NC could still be observed. Based on the *RC*-delay of the measurement setup, which was estimated to be around 30 ns, the shortest possible pulse length was limited to $\sim$100 ns. For this experiment, the sample with the thickest HZO (11.3 nm) and Al_2O_3 layer (4 nm) was used due to its lower overall capacitance. The obtained Q_d-V_{max} curve without hysteresis is shown in Fig. 6.15.

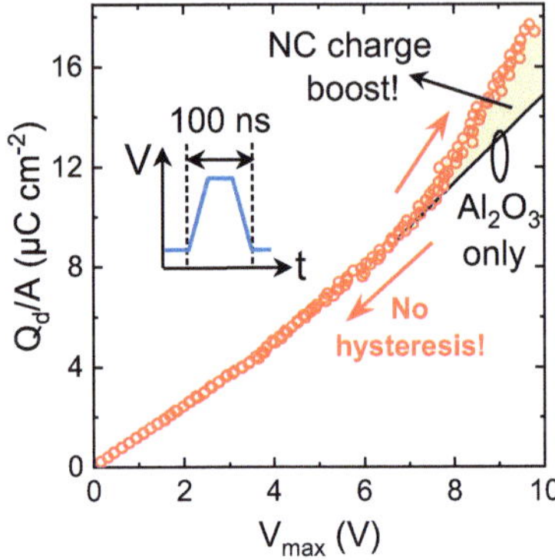

Figure 6.15: Shortest applied pulses of 100 ns still show hysteresis-free NC in the 11.3 nm HZO/4 nm Al_2O_3 sample. Applying even shorter pulses was not possible due to the *RC* delay of the measurement setup ($RC \approx 30$ ns). Adapted from Ref. [241]. © 2018 IEEE.

While this measurement indicates that the maximum speed of stabilized NC in HfO_2 based ferroelectrics is larger than 10 MHz frequency, it does not enable an assessment of the ultimate speed limit. However, this result disproves the argument that NC effects could only be used far below the MHz frequency range, as other authors have suggested based on comparisons to typical ferroelectric switching speeds in MFM capacitors, which involve a polarization hysteresis [46, 166].

Furthermore, the stabilized NC results for short pulses in Fig. 6.10 and 6.12 indicate that the observed NC effect might be an intrinsic effect, since the apparent coercive field in the extracted P-E_f 'S'-curves does not change with the ferroelectric thickness as would be expected for extrinsic stabilized NC, see Fig. 2.10a). However, the expected intrinsic coercive field for HfO_2 based ferroelectrics should be much higher than the observed $\sim$1.5 MV cm^{-1} based on first-principle calculations [98, 99]. Further research is necessary to determine if these NC effects in HfO_2 based ferroelectrics are intrinsic or extrinsic in nature.

6.1.5 Comparison to literature

While stabilized NC in ferroelectric/dielectric heterostructure and superlattice capacitors has been reported before, all previous results were based on perovskite ferroelectrics [7, 6, 239, 240]. For this reason, relatively thick ferroelectric and dielectric layers had to be used as shown in Fig. 6.16.

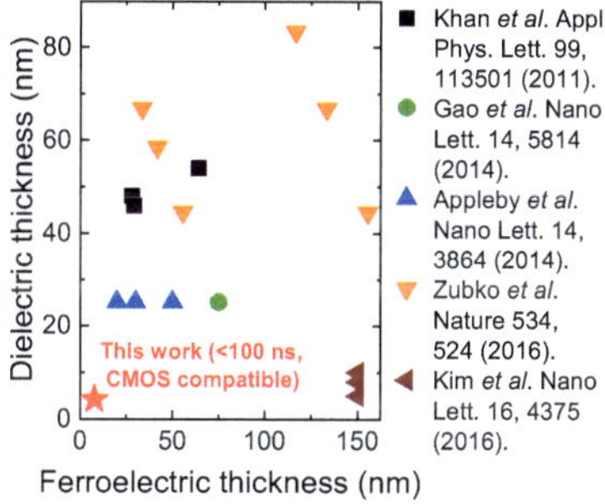

Figure 6.16: Comparison of hysteresis-free NC in ferroelectric/dielectric capacitors reported in literature. Besides this work, all previous reports applied perovskite-based ferroelectrics which are not CMOS compatible. Additionally, HZO is much more scalable compared to perovskite ferroelectrics. Adapted from Ref. [241]. © 2018 IEEE.

Indeed, only the superior scalability of HfO_2 based ferroelectrics has enabled the use of much thinner layers below 10 nm thickness as shown in this work. With the additional advantage of full CMOS process compatibility, so far HfO_2 based ferroelectrics seem to be the only realistic contender for implementations in nanoscale CMOS logic devices, also because they can be deposited on 3D structures, which will be necessary for advanced nodes where FinFETs or nanosheet channels are mandatory [255].

6.2 Ferroelectric/dielectric $Hf_{0.5}Zr_{0.5}O_2/Ta_2O_5$ capacitors

As mentioned previously, the reason why only a part of the NC region in the HZO/Al_2O_3 capacitors can be measured is the high electric field inside the Al_2O_3 layer, which leads to dielectric breakdown for the highest applied voltages. Starting from basic electrostatics, the electric field inside the dielectric layer in a ferroelectric/dielectric capacitor with interface charge σ_{IF} is given by

$$E_d = \frac{1}{\varepsilon_0 \left(\varepsilon_d + \varepsilon_b \frac{t_d}{t_f} \right)} \left[P_S - \sigma_{IF} + \frac{\varepsilon_0 \varepsilon_b}{t_f} V \right], \tag{6.4}$$

where V is the externally applied voltage to the capacitor. As can be seen from Eq. 6.4, increasing the thickness t_d and permittivity ε_d of the dielectric layer decreases the electric field for a given V and spontaneous polarization P_S. This indicates that using a thicker dielectric layer with higher permittivity would be beneficial to measure a larger fraction of the stabilized NC region in the ferroelectric. However, it is well known from experiment that the breakdown field strength of a dielectric is typically smaller the larger its permittivity [107]. Therefore, the most effective way to increase charge density $(P_S - \sigma_{IF})$ that the dielectric is able to withstand, is to increase the t_d/t_f ratio. However, this will lead to worse capacitance matching for a given permittivity. Therefore, increasing the permittivity of the dielectric will improve the capacitance matching for a higher

t_d/t_f ratio. For these reasons, samples with thicker dielectric Ta_2O_5 layers, instead of Al_2O_3, were fabricated together with a similar ferroelectric HZO layer. The relative permittivity of Ta_2O_5 was determined as $\varepsilon_d \approx 23.5$ from small-signal capacitance measurements on $TiN/Ta_2O_5/TiN$ capacitors. These permittivity values are roughly three times higher than for Al_2O_3. This section is based on the results published in Ref. [128].

6.2.1 Structural and standard ferroelectric characterization

The fabricated $TiN/HZO/Ta_2O_5/TiN$ heterostructure capacitors were characterized by high-resolution TEM and EELS as shown in Fig. 6.17.

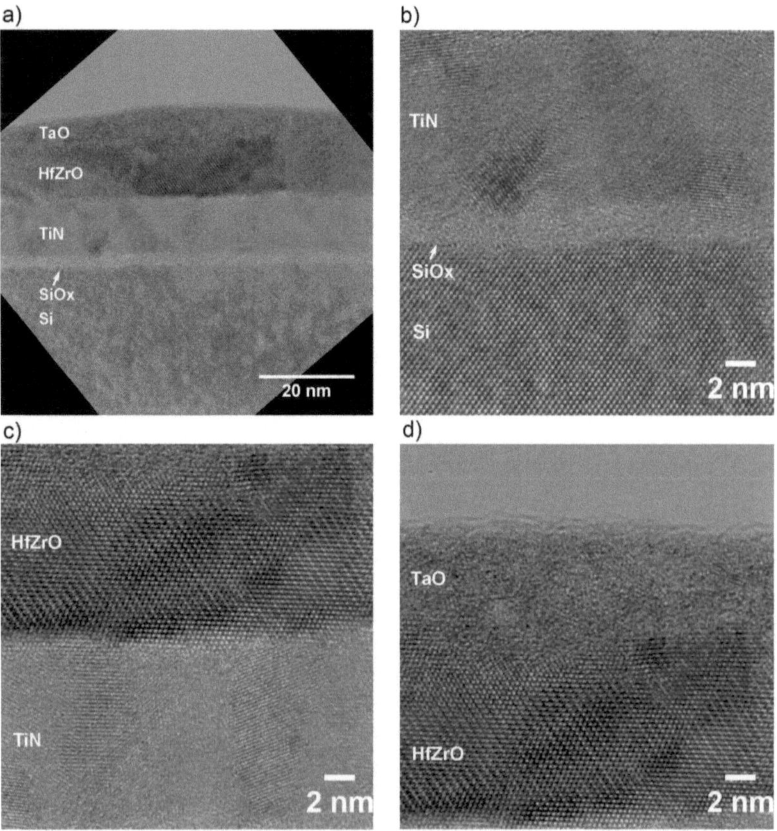

Figure 6.17: a) High-resolution TEM images of the $Ta_2O_5/HZO/TiN/SiO_x/Si$ heterostructure. b)–d), Magnified views of particular areas of a): b) the $TiN/SiO_x/Si$ interfaces, c) the HZO/TiN interface and d) the Ta_2O_5/HZO interface. Adapted from Ref. [128].

TEM cross-sections of the capacitor structure with magnifications of the $Si/SiO_x/TiN$ interface, the TiN/HZO interface and the HZO/Ta_2O_5 interface can be seen in Fig. 6.17. The polycrystalline nature of the HZO layer as well as the amorphous morphology of the Ta_2O_5 layer are visible as well. From the lower resolution TEM cross-section in Fig. 6.18a), also the thickness of the individual layers can be inferred. Furthermore, the EELS data in Fig. 6.18b) confirms the expected composition of the stacks, with only minor interdiffusion of Hf and Ta into the Ta_2O_5 and HZO layer, respectively.

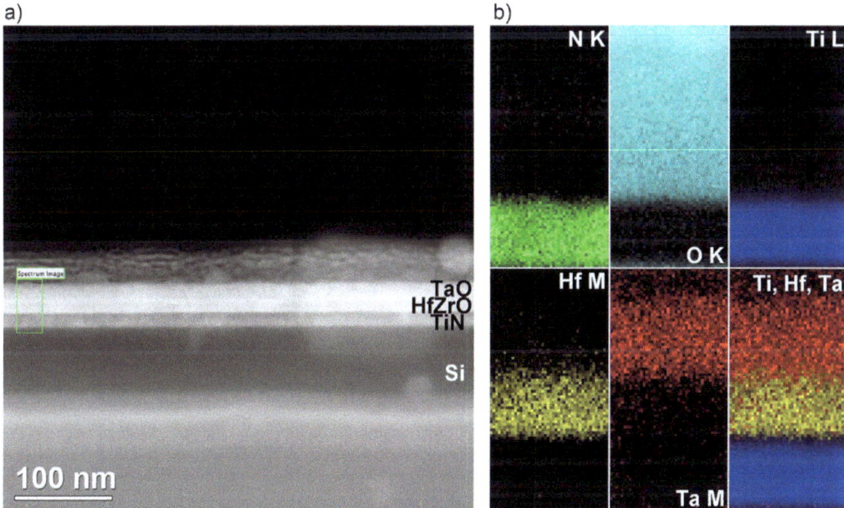

Figure 6.18: a) Low-magnification high-angle annular dark-field scanning TEM image of the $Ta_2O_5/HZO/TiN/SiO_x/Si$ heterostructure. The boxed region was analyzed by EELS. b) EELS maps corresponding to the boxed region in a) for the absorption edges N-K, O-K, Ti-L, Hf-M and Ta-M. The bottom right image was obtained by overlapping the Ti-L, Hf-M and Ta-M maps. Adapted from Ref. [128].

Standard electrical P-E_f and small-signal C-V characterization results measured on a TiN/HZO/TiN capacitor are shown in Fig. 6.19a,c). The remanent polarization of $\sim$17 µC cm^{-2} and coercive field of $\sim$1.2 MV cm^{-1}, which were obtained after 10^4 rectangular wake-up cycles at 100 kHz, are again consistent with previous reports [82]. Butterfly C-V curves were measured at a frequency of 10 kHz and allow the extraction of a relative background permittivity of 33. The GIXRD data of the HZO layer (Fig. 6.19b,d)) indicates mainly orthorhombic and tetragonal phase fractions in the film.

6.2.2 Pulsed charge-voltage measurement results

For the indirect measurement of hysteresis-free NC in the $TiN/HZO/Ta_2O_5/TiN$ heterostructure capacitor, the same pulses measurement technique as described in Section 6.1, which was adapted

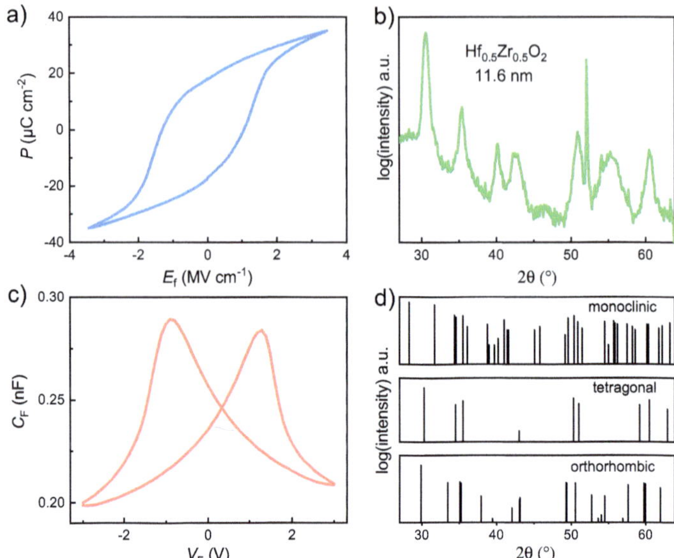

Figure 6.19: Electrical and physical characterization of the ferroelectric HZO layer. a) Ferroelectric polarization–electric field (P–E_f) hysteresis of the TiN/HZO/TiN capacitor measured by applying a 10-kHz triangular voltage waveform. b) GIXRD pattern of the HZO layer on Si. a.u., arbitrary units; 2θ is the angle between the incident X-rays and the detector. c, Small-signal capacitance–voltage (C–V) characteristics measured at 10 kHz on the same TiN/HZO/TiN capacitor as in a). d) Reference diffraction patterns for the monoclinic, tetragonal and orthorhombic phases in HfO_2. Adapted from Ref. [128].

from ref. [211]. Fig. 6.20a)-c) shows the schematic experimental setup, a TEM cross-section of the capacitor and the SAED pattern of the structure, respectively. The latter confirms orthorhombic diffraction spots from the ferroelectric HZO layer in agreement with GIXRD. Fig. 6.20d)-f) shows the applied voltage pulses, measured current and calculated charge, respectively. As discussed in Section 6.1, the maximum, discharged and residual charges are extracted from Fig. 6.20f) and plotted as a function of the voltage pulse amplitude in Fig. 6.20g). By using Eq. 6.2 and 6.3, the polarization and electric field inside the ferroelectric HZO layer can be calculated from the Q_d-V_{max} data in Fig. 6.20g), which is shown in Fig. 6.20h) together with the standard P-E_f characteristics of the TiN/HZO/TiN capacitor. The pulsed measurement data trace an 'S'-shaped curve as predicted by homogeneous LGD theory (red line, $\alpha = -4.6 \cdot 10^8$ m F^{-1} and $\beta = 9.8 \cdot 10^9$ m^5 C^{-2} F^{-1}). From the 'S'-shaped P-E_f curve in Fig. 6.20h), the ferroelectric double-well energy landscape can be calculated as

$$F_f(P) = F_0 + \int_{P_{min}}^{P_{max}} (E_f(P) - E_{bias})\mathrm{d}P, \tag{6.5}$$

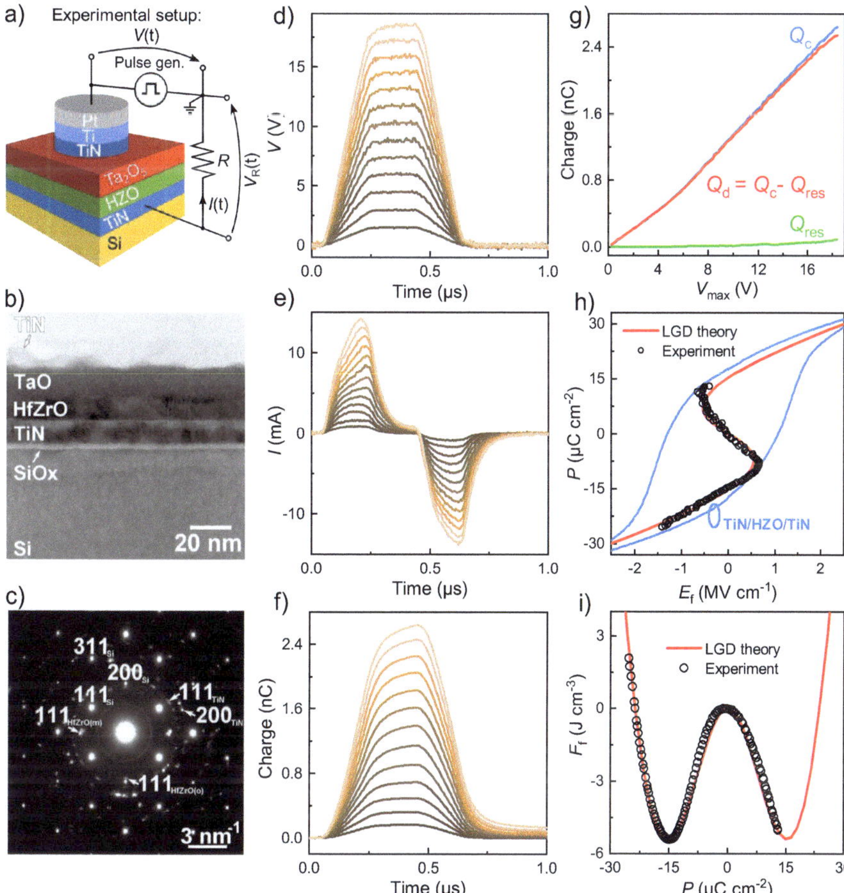

Figure 6.20: Electrical measurement of the ferroelectric double-well energy landscape. a) Schematic experimental set-up for the pulsed charge-voltage measurements. The voltage V is applied by a pulse generator, and both V and V_R are measured via an oscilloscope, where V_R is the voltage drop across the resistance R. b), TEM image of the ferroelectric/dielectric $TiN/Ta_2O_5/HZO/TiN$ heterostructure capacitor. c) SAED pattern of the heterostructure shown in b). TEM and the SAED pattern confirm a mixture of orthorhombic and monoclinic grains in the HZO layer and the amorphous nature of the Ta_2O_5 layer. d)-f), The applied voltage pulses, the measured current response and the integrated charge, respectively, as a function of time. g) The maximum charge Q_c and residual charge Q_{res}, together with their difference, $Q_d = Q_c - Q_{res}$; these are derived from f) as a function of the voltage across the capacitor (V_{max}) in d). h, The calculated polarization-electric field relationship of the HZO layer with a fit according to LGD theory. i) The result of integrating the P-E_f characteristics in h). The fit in i) uses the same LGD coefficients as in h). Adapted from Ref. [128].

where P_{max} and P_{min} are the maximum and minimum polarization values, E_{bias} is an internal bias field in the ferroelectric and the energy offset F_0 is chosen such that $F_f(P = 0) = 0$. Using E_{bias}

= 0.14 MV cm^{-1}, the measured double-well energy landscape in comparison to the theoretical Landau formula is shown in Fig. 6.20i). The excellent agreement between experiment and theory is surprising, since the HZO layer is a polycrystalline ferroelectric, which must have an inhomogeneous polarization distribution across the capacitor area but still behaves in this particular measurement as an almost ideal homogeneous 'Landau'-like ferroelectric.

However, the extracted energy barrier height from these experiments is roughly an order of magnitude smaller than predicted by first principles calculations [98, 99, 100], which is compatible with nucleation and growth mediated switching [217]. Nevertheless, such dissipative switching mechanisms should result in a polarization hysteresis. On the other hand, as mentioned in Section 6.1, the constant coercive field extracted in the pulsed NC measurements on the HZO/Al$_2$O$_3$ samples for different layer thicknesses suggests that these measured NC effects might be intrinsic, see Fig. 2.10. Further research is necessary to decide with certainty if the hysteresis-free NC observed in HfO$_2$ based ferroelectrics (see also [254]) is intrinsic or extrinsic.

To better understand, what happens inside the heterostructure capacitor during the measurement, the different electric field contributions inside the ferroelectric HZO layer and the dielectric Ta$_2$O$_5$ layer were calculated. Starting from the basic electrostatic boundary conditions

$$V_{max} = t_f E_f + t_d E_d \tag{6.6}$$

and

$$\varepsilon_0 \varepsilon_d E_d = \varepsilon_0 \varepsilon_b E_f + P_S - \sigma_{IF}, \tag{6.7}$$

the electric field inside the ferroelectric can be written as

$$E_f = \frac{1}{\varepsilon_0 \left(\varepsilon_b + \varepsilon_d \frac{t_f}{t_d} \right)} \left[-P_S + \sigma_{IF} + \frac{\varepsilon_0 \varepsilon_d}{t_d} V_{max} \right]. \tag{6.8}$$

Here,

$$E_{dep} = -\frac{P_S - \sigma_{IF}}{\varepsilon_0 \left(\varepsilon_b + \varepsilon_d \frac{t_f}{t_d} \right)} \tag{6.9}$$

is the depolarization field and the external field contribution due to the background permittivity of the ferroelectric is given by

$$E_{f,ext} = \frac{\varepsilon_d}{\varepsilon_b t_d + \varepsilon_d t_f} V_{max}. \tag{6.10}$$

The internal and external field contributions inside the dielectric layer can be directly obtained from Eq. 6.4 as

$$E_{d,int} = \frac{P_S - \sigma_{IF}}{\varepsilon_0 \left(\varepsilon_d + \varepsilon_b \frac{t_d}{t_f} \right)} \tag{6.11}$$

and

$$E_{d,ext} = \frac{\varepsilon_b}{\varepsilon_b t_d + \varepsilon_d t_f} V_{max}, \tag{6.12}$$

respectively. Using a simple LGD expression for the ferroelectric as $E_f = 2\alpha P_S + 4\beta P_S^3 + E_{bias}$, the external voltage can be calculated from the spontaneous polarization as

$$V_{max} = \left(t_f + t_d \frac{\varepsilon_b}{\varepsilon_d} \right) \left[2\alpha P_S + 4\beta P_S^3 + E_{bias} + \frac{P_S - \sigma_{IF}}{\varepsilon_0 \left(\varepsilon_b + \varepsilon_d \frac{t_f}{t_d} \right)} \right]. \tag{6.13}$$

Furthermore, the reversibly stored charge on the capacitor is then given by

$$Q_d = A[\varepsilon_0 \varepsilon_b (2\alpha P_S + 4\beta P_S^3 + E_{bias}) + P_S - \sigma_{IF}]. \tag{6.14}$$

Using Eq. 6.13 and 6.14, the measured Q_d-V_{max} characteristics can be well fitted as shown Fig. 6.21a). Additionally, Fig. 6.21b,c) shows the comparison between theory and experiment of the different electric field contritions inside the ferroelectric and dielectric layer, respectively.

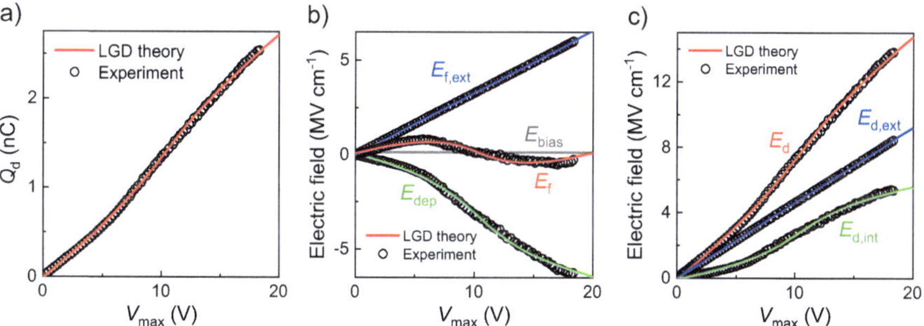

Figure 6.21: Comparison of theory and experiment of pulsed charge–voltage measurements on TiN/HZO/Ta$_2$O$_5$/TiN samples. a) Reversibly stored charge Q_d as a function of maximum voltage across the capacitor V_{max} as shown in Fig. 6.20; solid lines correspond to theoretical calculations using LGD theory. b) Calculated electric field contributions in the ferroelectric layer as a function of V_{max}. c) Calculated electric field contributions in the dielectric layer as a function of V_{max}. Adapted from Ref. [128].

As can be seen from Fig. 6.21, for the ferroelectric layer, the external and depolarization fields have opposite signs and largely compensate each other, such that the total field in the ferroelectric stays relatively low ($E_f < 1$ MV cm^{-1}) even for the highest applied voltages. In contrast, the external and internal field contributions in the dielectric have the same sign, which leads to an even higher total electric field E_d. Therefore, E_d reaches around 14 MV cm^{-1} at maximum applied voltage, thus leading to hard dielectric breakdown of the Ta$_2$O$_5$ layer.

6.2.3 Investigation of hysteresis

Similar to the previous experiments on the TiN/HZO/Al$_2$O$_3$/TiN samples discussed in Section 6.1, an investigation of the hysteresis behavior of the TiN/HZO/Ta$_2$O$_5$/TiN sample was carried out. Again, increasing and subsequent decreasing pulse amplitudes were used to obtain the Q_d-V_{max} characteristics, which are shown in Fig. 6.22a) with the corresponding P-E_f curve in Fig. 6.22b).

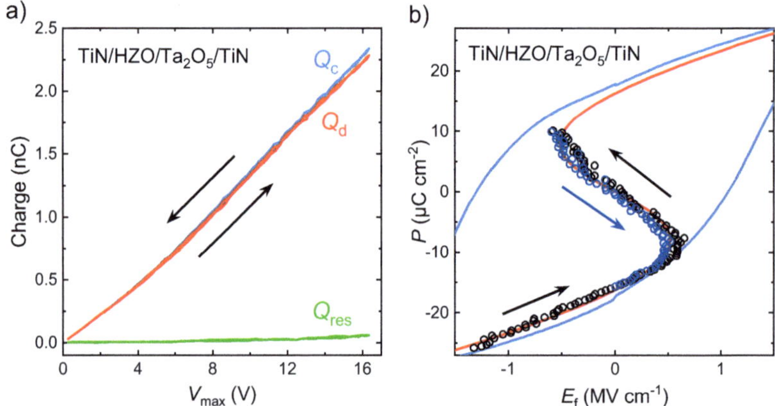

Figure 6.22: Confirmation of hysteresis-free operation and NC in TiN/HZO/Al$_2$O$_3$/TiN samples. a) Pulsed charge-voltage hysteresis measurement of a TiN/HZO/Ta$_2$O$_5$/TiN capacitor. Black arrows indicate sweep directions of increasing and decreasing V_{max}. b) 'S'-shaped P-E_f curve calculated from the data in a). Black data points and arrows correspond to the increasing V_{max} sweep direction, while blue data points and arrow indicate the decreasing V_{max} sweep direction. The red line corresponds to LGD theory, and the blue line shows the P-E_f hysteresis measured on a TiN/HZO/TiN sample. Adapted from Ref. [128].

For a maximum applied voltage amplitude of ~16 V, only negligible hysteresis is observed, which is consistent with only small Q_{res} indicating almost no leakage or charge trapping. However, as was noted by Kim et al. [211], the pulse length can have an important effect on the observed hysteresis. It was observed, that longer pulse times lead to an emergence of hysteresis, due to an increase in charge trapping at the ferroelectric/dielectric interface [211]. To investigate this behavior in the TiN/HZO/Ta$_2$O$_5$/TiN sample, the pulse time was increased from originally 400 ns up to 400 µs. The resulting Q_d-V_{max} and P-E_f curves are shown in Fig. 6.23a) and 6.23b), respectively.

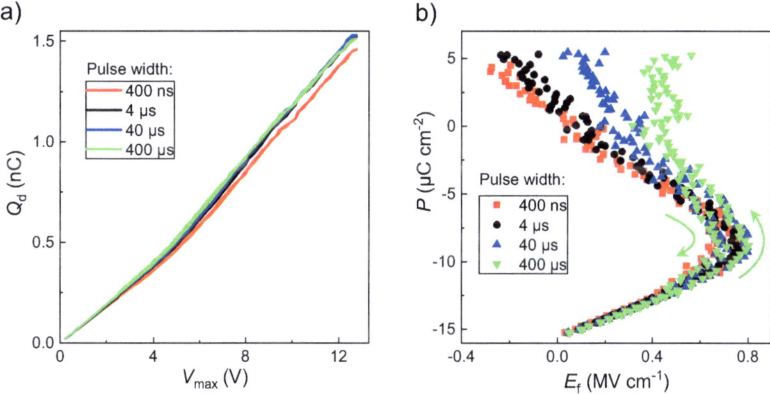

Figure 6.23: Investigation of the transition from hysteresis-free NC to hysteretic switching. a) Pulsed charge-voltage hysteresis measurement of a TiN/HZO/Ta$_2$O$_5$/TiN capacitor (same as in Fig. 6.20) for different pulse widths. b) P-E_f curves calculated from the data in a). Green arrows indicate hysteresis when applying first increasing and then decreasing voltage pulses V_{max}. Adapted from Ref. [128].

While the hysteresis is generally increasing with increasing pulse width, it stays relatively small even for the 400 µs pulse widths. However, a progressive shift of the P-E_f curves towards higher fields at high polarization can be observed. This might be caused by the continuous injection of charges into the ferroelectric/dielectric interface, which changes the electrostatic boundary conditions, which were assumed constant in the extraction of Fig. 6.23b). The frequency dispersion of the permittivity of the Ta$_2$O$_5$ layer was considered in these calculations. For 400 ns, 4 µs, 40 µs and 400 µs, the permittivity was measured to be 23.5, 24.7, 25.5 and 26.1, respectively.

6.2.4 Dynamic simulation of hysteresis-free negative capacitance

While it was argued in the previous section, that the overall extracted Q_d-V_{max} curve is hysteresis-free when extracting the charges after each pulse only, the question of hysteresis in the *transient* characteristics during each pulse remains. To understand the time-resolved NC behavior in the ferroelectric HZO layer, a dynamic circuit model was developed. While Fig. 6.24a) depicts the schematic experimental setup used before, Fig. 6.24b) shows the corresponding equivalent electric circuit, which was simulated. Importantly, a leakage current path through the capacitor was added as indicated by the parallel resistance R_L. The leakage current is modeled using a simple Fowler-Nordheim approximation in the form of $I_L = k_1(V_D + V_F)^2 \exp(-k_2/|V_D + V_F|)$, where k_1 and k_2 are fitting constants. The parasitic series resistance of the setup and sample is included in R. The ferroelectric is modeled by the same Landau parameters as used in Fig. 6.20h). No intrinsic ferroelectric hysteresis was assumed, corresponding to an internal resistance $\rho = 0$ in the Landau-Khalatnikov equation 2.21. The only values that were fitted to the experimental data were

$k_1 = 1.85 \, \mu\text{A} \, \text{V}^{-2}$, $k_2 = 11 \, \text{V}$ and $R = 350 \, \Omega$. All other parameters were taken from independent measurements. The measured applied voltage V was directly taken as the input for the simulations.

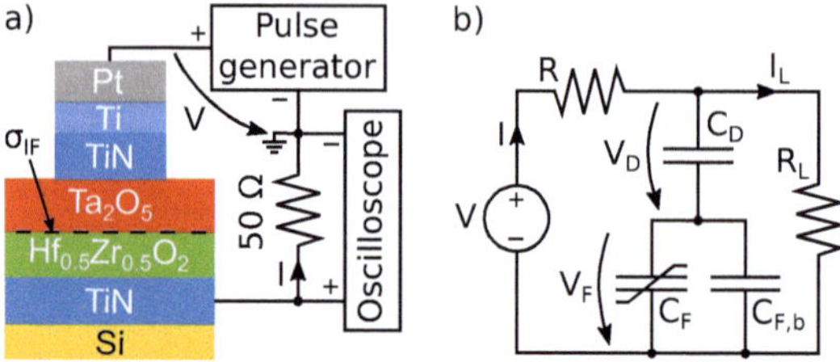

Figure 6.24: a) Sample structure and experimental setup for fast pulsed voltage NC measurements. b) Equivalent circuit model for dynamic NC simulations of the setup in a). Adapted from Ref. [256]. © 2019 IEEE.

Fig. 6.25b,c) shows the dynamic simulation results for the current I and the charge Q in comparison to four measurement curves from Fig. 6.20e,f). The dynamic simulation and measurement curves are in good agreement for the complete voltage range and time domain.

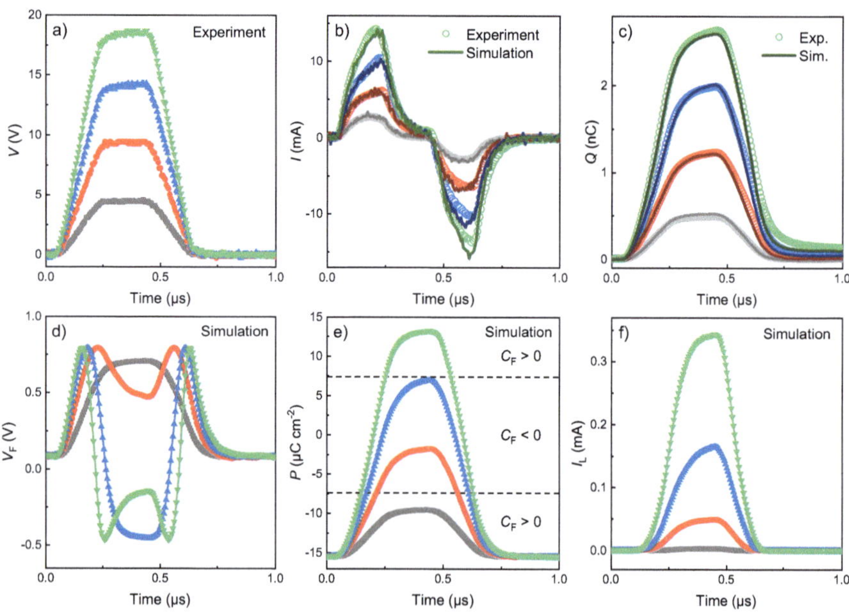

Figure 6.25: a) Experimental voltage pulses used for the simulation. Comparison of simulated and experimental (Ref. [242]) b) transient currents and c) integrated charges from b). Simulated transient d) voltage across the ferroelectric V_F, e) ferroelectric polarization P, and f) leakage current through the capacitor I_L. Adapted from Ref. [256]. © 2019 IEEE.

The only minor deviations are observed during the discharging process for higher voltages, which might be related to the simplified Fowler-Nordheim leakage model. These results suggest, that the NC observed in the TiN/HZO/Ta$_2$O$_5$/TiN capacitors is indeed hysteresis-free, even in its transient characteristics. This also indicates, that the speed limit of NC in these ferroelectric HZO films is much smaller than the RC-delay of the measurement setup.

From the simulation, the internal circuit variables like V_F, P and I_L are also obtained, which can be plotted as a function of time as shown in Fig. 6.25d)-f). From the transient V_F curves, the time frames of NC in the ferroelectric can be directly seen. For the gray curve, with the lowest voltage pulse amplitude, the HZO layer stays in the positive capacitance regime for the whole time, since V_F always changes in the same direction as the charge Q. However, for the red curve, which corresponds to around 9 V amplitude, as soon as V_F reaches $\sim$0.8 V, the ferroelectric enters the NC region and V_F changes opposite to the change in Q. The same can be observed for the blue V_F curve, where V_F even takes on negative values for high enough applied voltages V. For the highest amplitude (green V_F curve), the change in charge is so high, that the ferroelectric even leaves the NC region for the highest voltages and enters the second positive capacitance region. This can also be seen from the transient polarization in Fig. 6.25e), where the different positive and negative capacitance regions are indicated by the dashed horizontal lines. The simulated leakage current I_L in Fig. 6.25f) is shown to be relatively low compared to the overall current I.

To compare the final results to the experimental data, Q_d, Q_c and Q_{res} as a function of V_{max} were extracted from the simulation. Fig. 6.26 shows the corresponding graphs comparing the simulation results and the measured data, which is in excellent agreement.

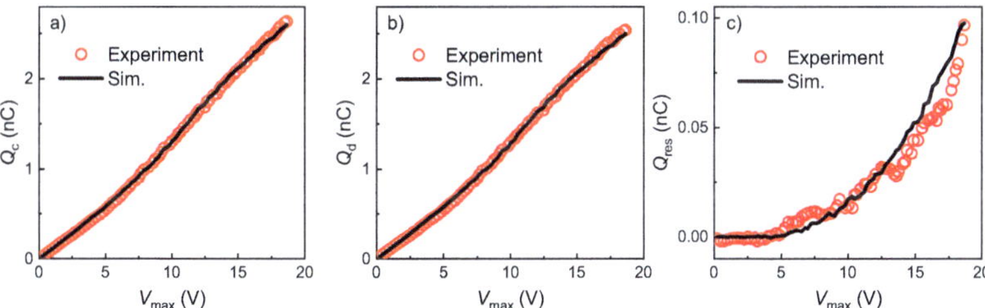

Figure 6.26: Comparison of experimental data (Ref. [128]) and simulations of the integrated charges as a function of the maximum pulse height: a) Maximum charge Q_c, b) reversibly stored charge Q_d and c) residual charge Q_{res}. Adapted from Ref. [256]. © 2019 IEEE.

6.3 Negative capacitance electrostatic supercapacitors

The relatively large voltage (> 5 V) needed to observe NC in these ferroelectric/dielectric heterostructure capacitors with negative interface charges makes the direct application in low power electronics difficult, since operating voltages below 1 V are needed. However, NC appearing at higher voltages can have advantages for applications beyond low power electronics. In the following, the use of NC in ferroelectric/dielectric structures as described in Sections 6.1 and 6.2 for energy storage applications is proposed. While electrostatic energy in regular capacitors can be stored with highest power densities, the energy storage density is much lower than for example that of batteries of fuel cells [257]. However, so-called supercapacitors combine the advantages of very high power densities with much higher energy densities than regular capacitors. In current electrochemical supercapacitors, this is achieved by the electric double-layer capacitance and pseudocapacitance effects [258, 259].

Recently, purely electrostatic solid-state supercapacitors based on highly polarizable materials like ferroelectrics and antiferroelectrics [260, 261, 262] have received increasing attention. However, these types of supercapacitors have reduced energy storage efficiency due to hysteretic polarization switching. Therefore, utilizing hysteresis-free NC in (anti)ferroelectric materials, highly-efficient supercapacitors might be built. First, a general theory for this kind of NC supercapacitor will be derived, which establishes the theoretical limits of this device. Secondly, experimental characteristics of different $TiN/HZO/Ta_2O_5/TiN$ and $TiN/HZO/Al_2O_3/TiN$ capacitors are investigated with a focus on the energy storage characteristics. Lastly, reliability measurements are carried out and the overall results are compared to previous reports on electrostatic solid-state supercapacitors in literature. This section is based on the results published in Ref. [242].

6.3.1 Concept and theory

The energy density stored in the electric field E of a dielectric material is given by

$$w = \int_0^D E(D)\mathrm{d}D. \tag{6.15}$$

In general, the relationship between E and the electric displacement field D can be non-linear. For a parallel-plate capacitor with an applied voltage V between the electrodes and the charge Q on the electrodes, the total electrostatic energy (neglecting fringing fields) can be written as

$$W = \int_0^Q V(Q)\mathrm{d}Q. \tag{6.16}$$

The stored energy in a capacitor can thus be obtained graphically as the area above its Q-V curve. In the case of a linear dielectric material, $D = \varepsilon_0\varepsilon_d E$, the capacitance $C = Q/V$ is constant. This means that the stored energy is given by the well-known expressions $W = CV^2/2$ or $w = \varepsilon_0\varepsilon_d E^2/2$, corresponding to the area of the green triangle in Fig. 6.27a). When thinking about NC on the other hand, in any passive device, the NC region must be bounded by regions of positive capacitance, which generally leads to an 'S'-shaped charge-voltage relationship as shown in Fig. 6.27b). When

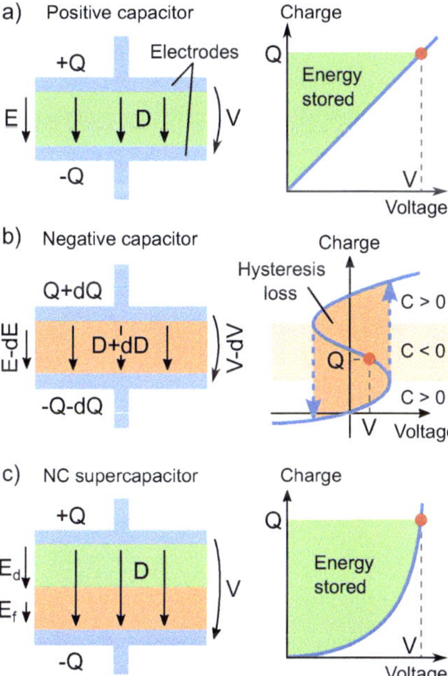

Figure 6.27: Combining positive and negative capacitance to enhance energy storage. a) Positive capacitor with charge Q, voltage V, electric field E, and displacement field D. Energy stored in a capacitor is given by the area (green) above the Q-V curve. b) Capacitor with negative differential capacitance $dQ/dV < 0$ due to $dD/dE < 0$. Exemplary Q-V characteristic with NC region. In isolation, the NC region is unstable, and hysteresis is observed (dashed blue arrows) leading to an energy loss (orange area). c) Combining NC and positive capacitance to create an NC supercapacitor with improved energy storage (green area) and without hysteresis. Adapted from Ref. [242].

the voltage is swept and the charge is measured, a jump in charge will be measured as indicated by the dashed arrows. The electrical energy that is turned into heat due to hysteresis losses can be calculated as the difference between the charging energy W_c and the released energy during discharging W_d as

$$W_{loss} = W_c - W_d = \int_0^Q V(Q)\mathrm{d}Q - \int_Q^0 V(Q)\mathrm{d}Q. \tag{6.17}$$

By dividing the total energies W_c, W_d and W_{loss} by the capacitor volume, the respective average energy densities $\langle w_c \rangle$, $\langle w_d \rangle$ and $\langle w_{loss} \rangle$ are obtained. Furthermore, the energy storage efficiency η can be defined as

$$\eta = \frac{W_d}{W_c} = \frac{W_d}{W_d + W_{loss}}. \tag{6.18}$$

When a positive and negative capacitance layer are combined as shown in Fig. 6.27c), the green area under the Q-V curve, i.e. the stored energy, can be increased due to the non-linearity induced by the NC layer. In this way, the stored energy can be increased, even for identical charge and voltage compared to a linear capacitor. By matching the capacitances of both layers, the overall system shows no hysteresis, leading to a theoretical efficiency of $\eta = 1$.

In the following, based on a general and simplified theory, the concept of the NC supercapacitor will be described. Without specifically referring to ferroelectrics yet, an 'S'-shaped Q-V characteristic as shown in Fig. 6.27b) can be described by a 3rd order polynomial as

$$V = a(Q - \Delta Q) + b(Q - \Delta Q)^3, \tag{6.19}$$

where $a < 0$, $b > 0$ and $\Delta Q = \sqrt{-a/b}$. Since a linear positive capacitor with capacitance C obeys $V = Q/C$, the series combination between the NC layer and the positive capacitance layer can be written as

$$V = \left(\frac{1}{C} + a\right) Q - a\Delta Q + b(Q - \Delta Q)^3. \tag{6.20}$$

From Eq. 6.20 it can be seen that only under the condition $1/C > -a$, the total capacitance $C = dQ/dV$ is positive for all Q. That means, that if this condition is fulfilled, the combined capacitor will have no hysteresis, enabling $\eta = 1$. This is identical to the original proposal by Salahuddin and Datta for hysteresis-free NC transistors [4], see also Eq. 2.30. Furthermore, if $1/C = -a$ (i.e. $aC = -1$), the capacitances of both layers are ideally matched, yielding the largest voltage amplification and overall capacitance enhancement. Using Eq. 6.16 and 6.20, the total stored energy in the two-layer capacitor can be written as

$$W = \left(\frac{1}{2C} - a\right) Q^2 - b\Delta Q Q^3 + \frac{b}{4}Q^4. \tag{6.21}$$

To calculate the enhancement of the total stored energy of the two-layer capacitor W compared to the stored energy $W_0 = Q_0^2/2C$ in a single-layer positive capacitor with capacitance C at identical voltage V, an expression of the charge Q_0 on the positive capacitor by itself is needed. Since W and W_0 should be compared at the same voltage, one has to solve

$$\frac{Q_0}{C} = \left(\frac{1}{C} + a\right) Q - a\Delta Q + b(Q - \Delta Q)^3, \tag{6.22}$$

for Q_0, which can then be inserted into $W_0 = Q_0^2/2C$ yielding

$$W_0 = \frac{1}{2C}\left[(1 + aC)^2 Q - aC\Delta Q + bC(Q - \Delta Q)^3\right]^2. \tag{6.23}$$

By normalizing W_0 to W, the energy storage enhancement W/W_0 can then be defined as

$$\frac{W}{W_0} = \frac{(1-2aC)Q^2 - 2bC\Delta QQ^3 + 0.5bCQ^4}{[(1+aC)^2Q - aC\Delta Q + bC(Q-\Delta Q)^3]^2}. \tag{6.24}$$

For ideal capacitance matching Eq. 6.24 simplifies to

$$\left.\frac{W}{W_0}\right|_{aC=-1} = \frac{6\Delta Q^4 - 4\Delta Q^3 Q + \Delta Q^2 Q^2}{2(3\Delta Q^2 - 3\Delta QQ + Q^2)^2}. \tag{6.25}$$

While the analytical expression of the maximum of Eq. 6.25 with respect to Q would be rather complicated, it was found that the maximum appears at roughly $Q_x = 1.4\Delta Q$. This leads to a maximum energy storage enhancement of

$$\max\left(\left.\frac{W}{W_0}\right|_{aC=-1}\right) = \frac{W}{W_0}(Q \approx 1.4\Delta Q) \approx 2.043. \tag{6.26}$$

This means, that for ideal capacitance matching ($aC = -1$), the stored energy in the NC supercapacitor can in theory be twice as large compared to a single-layer positive capacitor at identical voltage $V_x = V(Q = Q_x) \approx 1.07a\Delta Q$. Any energy storage enhancement $W/W_0 > 1$ would be impossible when only positive capacitance layers are combined, since the total capacitance will be always smaller than each of the individual capacitances, and $W_0 = CV^2/2$. In a next step, it will be investigated where the energy is actually stored in the NC supercapacitor, in the positive or in the negative capacitance layer? The total stored energy W must always be the sum of the energy stored in the positive capacitance layer W_{PC} and the energy stored in the NC layer $W_{NC} = W - W_{PC}$. Fig. 6.28 shows the stored energies normalized to W_x, which is the total energy at maximum energy storage enhancement, as a function of the normalized voltage and charge for ideal capacitance matching.

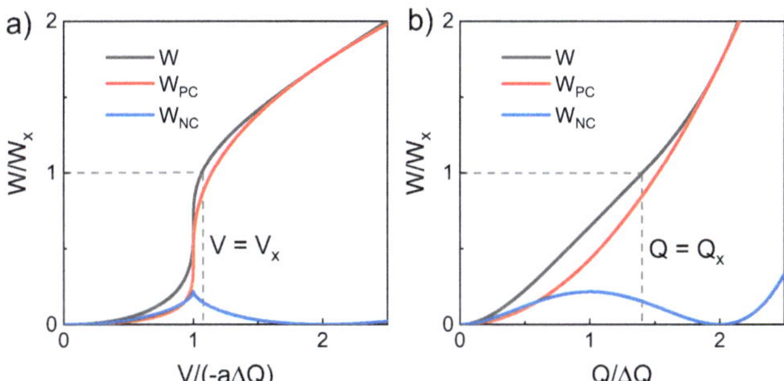

Figure 6.28: Total stored energy W, energy of the positive capacitance layer W_{PC} and energy of the NC layer W_{NC} as a function of a) the voltage V and b) the charge Q, normalized to the remanent charge ΔQ. W_x is the total energy at maximum energy storage enhancement. Adapted from Ref. [242].

As can be seen from Fig. 6.28, for maximum energy storage enhancement ($W/W_x = 1$), most of the energy is actually stored in the positive capacitance layer and not in the NC layer. Indeed, only about 15 % of W is stored in the NC layer under these conditions. For even larger voltages $V > V_x$ or charges $Q > Q_x$ the fraction of W_{NC}/W is further reduced. The reason for this is that the voltage drop across the NC layer in the NC region is much smaller than the voltage drop across the positive capacitance layer, due to the voltage amplification effect. Furthermore, the behavior in Fig. 6.28 is independent of the parameter b.

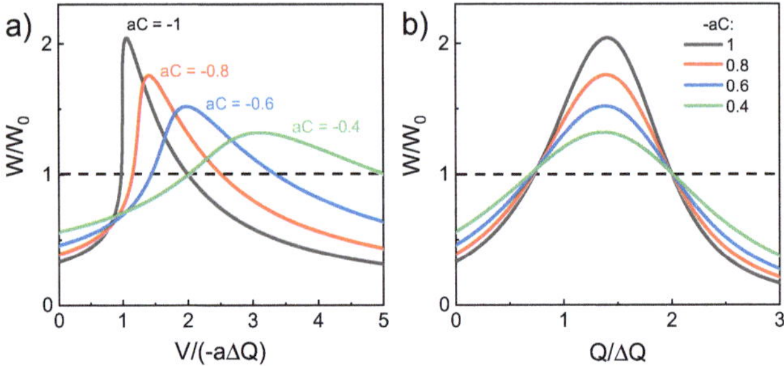

Figure 6.29: Energy storage enhancement W/W_0 as a function of the voltage V (a)) and the charge Q (b)), normalized to the remanent charge ΔQ for different capacitance matching conditions. Adapted from Ref. [242].

The influence of the capacitance matching on the energy storage enhancement as a function of the normalized voltage $V/(-a\Delta Q)$ and charge $Q/\Delta Q$ is shown in Fig. 6.29a) and b), respectively. For better capacitance matching (aC closer to -1), the maximum capacitance enhancement increases. Furthermore, the voltage range for capacitance enhancement $W/W_0 > 1$ decreases for better capacitance matching. In contrast, the charge range for $W/W_0 > 1$ is independent of the capacitance matching as can be seen from Fig. 6.29b). This also means that the charge Q_x at maximum energy storage enhancement changes only slightly with aC, which is also shown in Fig. 6.30a).

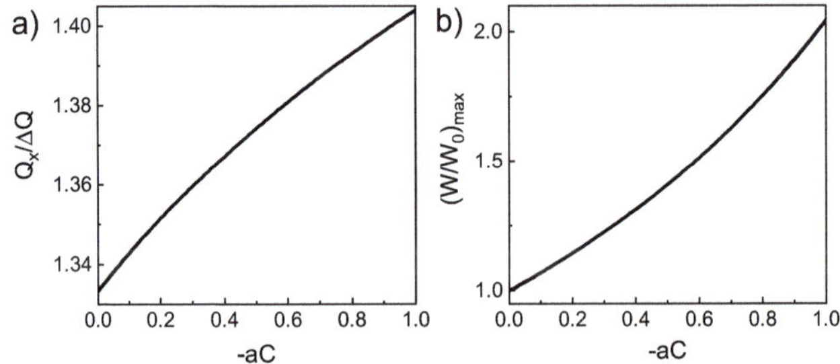

Figure 6.30: a) Charge at maximum energy storage enhancement normalized to the remanent charge ΔQ as a function of capacitance matching. b) Maximum energy storage enhancement as a function of capacitance matching. $aC = -1$ means perfect capacitance matching, while $aC = 0$ indicates worst matching. Adapted from Ref. [242].

The maximum energy storage enhancement as a function of $-aC$ is depicted in Fig. 6.30b). In a first-order approximation, the maximum storage enhancement is then given by

$$\left.\frac{W}{W_0}\right|_{aC=-1} \approx 1 - aC, \tag{6.27}$$

which means that for complete capacitance mismatch ($aC \approx 0$), the storage enhancement $W/W_0 \approx 1$ and for ideal matching ($aC = -1$) it is roughly two. The voltage V_x at which the maximum energy storage enhancement occurs, is approximately given by

$$V_x \approx \left(\frac{4}{3C} + \frac{1}{4}a\right)\Delta Q. \tag{6.28}$$

Lastly, the fraction of the energy that is stored in the NC layer also depends on the capacitance matching. In Fig. 6.31 it is shown that for non-ideal capacitance matching, the fraction of the stored energy in the NC layer is even further reduced and approaches zero for complete mismatch ($aC \approx 0$). This means that in general, while the NC layer serves a voltage amplification layer to engineer the overall capacitance non-linearity, the positive capacitance layer is the main storage layer in the NC supercapacitor. In addition, the use of the NC layer also leads to reduced leakage currents through the capacitor and an improved breakdown field strength of the positive capacitance layer, increasing the practical energy storage density even further [263].

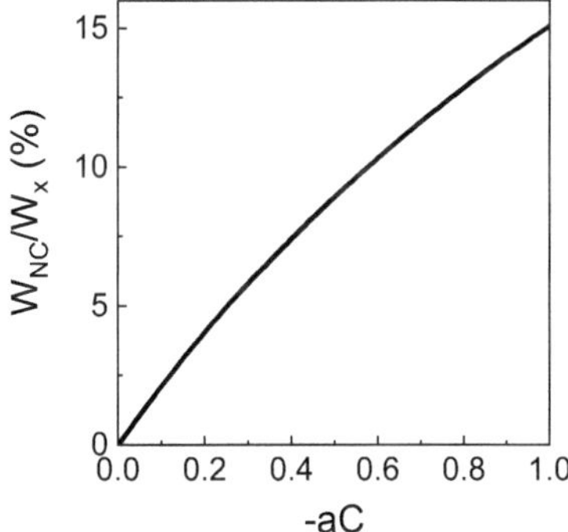

Figure 6.31: Fraction of the stored energy in the NC layer at maximum energy storage enhancement as a function of capacitance matching. $aC = -1$ means perfect capacitance matching, while $aC = 0$ indicates worst matching. Adapted from Ref. [242].

While so far, only the general concept of the NC supercapacitor was described, in the following, the concrete realization of a ferroelectric/dielectric supercapacitor with fixed interface charges will be presented. For simplicity, it is assumed here that the spontaneous ferroelectric polarization is equal to the ferroelectric displacement field, i.e. $\varepsilon_b = 0$. The Helmholtz free energy densities of the dielectric F_d and the ferroelectric F_f can then be written as

$$F_d = \frac{1}{2\varepsilon_0\varepsilon_d}D_d^2,\tag{6.29}$$

$$F_f = \alpha P^2 + \beta P^4.\tag{6.30}$$

The electric fields inside the layers can then be obtained from

$$E_d = \frac{\mathrm{d}F_d}{\mathrm{d}D_d} = \frac{D_d}{\varepsilon_0\varepsilon_d},\tag{6.31}$$

$$E_f = \frac{\mathrm{d}F_f}{\mathrm{d}P} = 2\alpha P + 4\beta P^3.\tag{6.32}$$

The capacitances per area A are then given by

$$\frac{1}{C_d} = t_d\frac{\mathrm{d}E_d}{\mathrm{d}D_d} = \frac{t_d}{\varepsilon_0\varepsilon_d},\tag{6.33}$$

$$\frac{1}{C_f} = t_f\frac{\mathrm{d}E_f}{\mathrm{d}P} = t_f(2\alpha + 12\beta P^2).\tag{6.34}$$

Fig. 6.32a) and b) show the free energy, displacement field and capacitance of a short-circuited single-layer dielectric and ferroelectric capacitor, respectively. The blue circles indicate the thermodynamic equilibrium states. While the dielectric is in equilibrium at $D_d = 0$, the ferroelectric shows a non-zero spontaneous polarization, which is compensated by the free electrons in the metal electrodes. Furthermore, the dielectric capacitance C_d is constant with D_d, while the ferroelectric capacitance C_f strongly changes with P and is negative for low P.

When combining the ferroelectric and dielectric layers into a ferroelectric/dielectric heterostructure capacitor with an interface charge σ_{IF}, the electrostatic boundary conditions

$$V = t_f E_f + t_d E_d\tag{6.35}$$

and

$$D_d = \varepsilon_0\varepsilon_d E_d = P - \sigma_{IF}\tag{6.36}$$

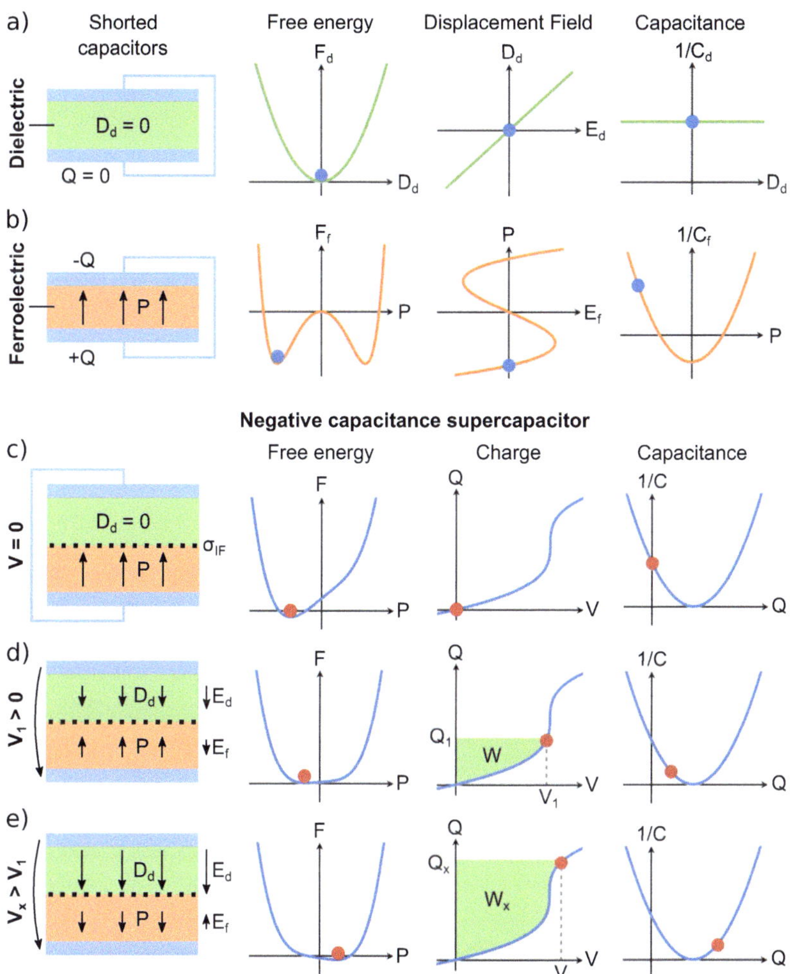

Figure 6.32: Realizing NC supercapacitors using a ferroelectric/dielectric stack. a) Free energy F_d, displacement field D_d, and inverse capacitance C_d^{-1} of a dielectric capacitor with shorted electrodes. b) The same representation for a ferroelectric capacitor with shorted electrodes and spontaneous polarization P. c) Proposed NC supercapacitor consisting of a ferroelectric/dielectric stack with interfacial charge σ_{IF}, which stabilizes P at zero voltage. d) NC supercapacitor under application of a positive voltage $V_1 > 0$. e) NC supercapacitor under application of a higher positive voltage V_x. Due to the NC in the ferroelectric layer, the stored energy (green area) is significantly enhanced compared to a regular dielectric capacitor. Adapted from Ref. [242].

must be fulfilled. The total free energy per capacitor area is then given by

$$F = t_f F_f + t_d F_d = t_f(\alpha P^2 + \beta P^4) + \frac{t_d}{2\varepsilon_0\varepsilon_d}(P - \sigma_{IF})^2. \tag{6.37}$$

Furthermore, the voltage applied to the electrodes can be written as

$$V = t_f(2\alpha P + 4\beta P^3) + t_d \frac{P - \sigma_{IF}}{\varepsilon_0\varepsilon_d}. \tag{6.38}$$

Here one can see, when comparing Eq. 6.38 to Eq. 6.20, that $Q = A(P - \sigma_{IF})$ and $\Delta Q = -A\sigma_{IF}$. Furthermore, $a = 2\alpha t_f/A$ and $b = 4\beta t_f/A^3$ as well as $C = \varepsilon_0\varepsilon_d A/t_d$. Eq. 6.38 with $\sigma_{IF} = -\sqrt{-\alpha/2\beta}$ and $1/C_d = 2\alpha t_f$ gives rise to the non-linear Q-V characteristics shown in Fig. 6.32c)-e). Furthermore, the total capacitance C can be calculated as

$$\frac{A}{C} = \frac{t_d}{\varepsilon_0\varepsilon_d} + t_f\left(2\alpha + 12\beta\left(\frac{Q}{A} + \sigma_{IF}\right)^2\right). \tag{6.39}$$

Note that, in Fig. 6.32c)-e), the total capacitance is always positive and approaches infinity in the NC region. Now, the operation principle of the ferroelectric/dielectric supercapacitor can be easily seen by looking at the free energy, charge and capacitance curves for different applied voltages. For the short-circuit condition $V = 0$ in Fig. 6.32c), the ferroelectric polarization points towards the ferroelectric/dielectric interface, because the interfacial charge σ_{IF} is assumed to be negative in this case, thus completely screening the depolarization field. The capacitance in this region is rather low. Now when a positive voltage $V > 0$ is applied (see Fig. 6.32d)), the ferroelectric polarization as well as the charge on the capacitor increases slowly, leading to moderate positive electric fields in both layers and increasing the stored energy W. The overall capacitance is increasing. Now, if the voltage is further increased towards $V = V_x$ (see Fig. 6.32e)), the ferroelectric enters the NC region leading to an amplification of the voltage drop across the dielectric. Both polarization and charge are rapidly increasing with only a small change of the voltage, due to the very large total capacitance. The stored energy is drastically increased towards $W = W_x$.

In the following the previous theoretical, concept will be demonstrated experimentally in TiN/H-ZO/ Al$_2$O$_3$/TiN and TiN/HZO/Ta$_2$O$_5$/TiN capacitors, which were already described in Section 6.1 and 6.2, respectively.

6.3.2 Experimental demonstration

First, the previously discussed data from Fig. 6.20g) is analyzed with respect to the energy storage properties, as shown in Fig. 6.33. The region of NC is indicated by the colored background in Fig. 6.33b). The calculated energy densities during charging, discharging and their difference as a function of the applied voltage are shown in Fig. 6.33c). As can be also seen from the calculated efficiency in Fig. 6.33d), the lost energy during charging and discharging is very small leading to efficiencies of 94.4 % even for the highest discharged energy densities of 109 J cm^{-3}.

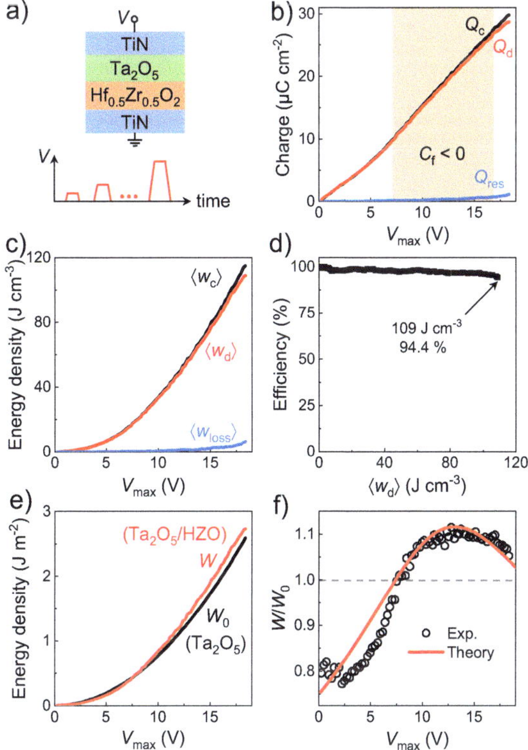

Figure 6.33: Experimental demonstration of an NC supercapacitor with high efficiency and energy density. a) NC supercapacitor sample structure using a Ta$_2$O$_5$ dielectric and a HZO ferroelectric layer. To measure the stored charge, short (400 ns) voltage pulses are applied at the top electrode. b) Measured charges as a function of the applied voltage during charging (Q_c) and discharging (Q_d). The difference is defined as $Q_{res} = Q_c - Q_d$, as reported in ref. [242]. c) Similarly, the average energy densities during charging ($\langle w_c \rangle$) and discharging ($\langle w_d \rangle$) as well as their difference ($\langle w_{loss} \rangle$) are shown as a function of V. d) Energy storage efficiency η as a function of $\langle w_d \rangle$. e) Comparison of discharged energy densities for the Ta$_2$O$_5$/HZO stack (W) and the Ta$_2$O$_5$ layer alone (W_0, extrapolated based on small-signal permittivity). f) Comparison of experimental and theoretical W/W_0. Adapted from Ref. [242].

When comparing the stored energy W per unit area to the theoretical limit W_0 of the Ta$_2$O$_5$ layer alone (based on the measured relative permittivity of 23.5), it can be seen that $W/W_0 > 1$ starting in the NC region (see Fig. 6.33e,f)). Using the fitted Landau parameters α and β from Section 6.2, the experimental data in Fig. 6.33f) agrees well with the theoretical curve (Eq. 6.20 and 6.24).

For electrostatic energy storage applications, reliability aspects are of critical importance. Therefore, the reliability of the TiN/HZO/Ta$_2$O$_5$/TiN samples was investigated with respect to frequency, cycling and temperature stability. Fig. 6.34a,c) shows the change of small-signal capacitance C, loss factor $\tan(\delta)$ and average discharged energy storage density $\langle w_d \rangle$ with frequency. In the investigated frequency range (1 kHz to 1 MHz), the discharged energy storage density does not change significantly for the applied voltage amplitude of 12.7 V.

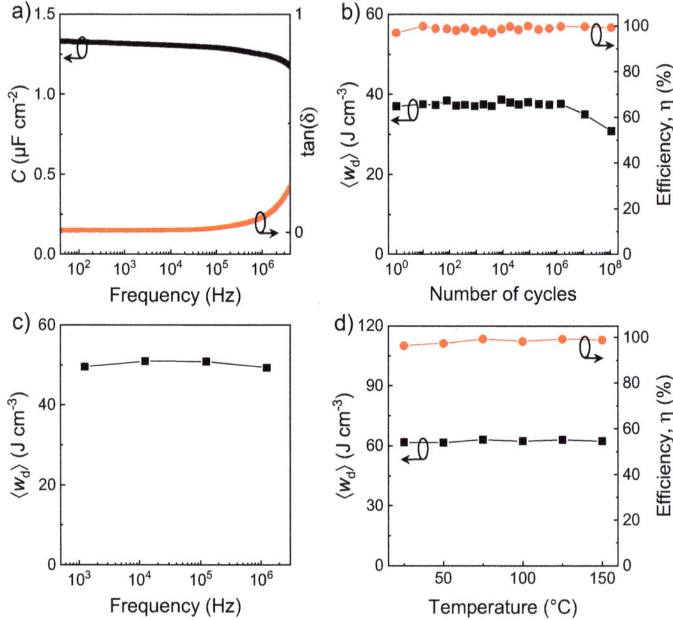

Figure 6.34: Frequency, cycling, and temperature stability of the NC supercapacitor. a) Capacitance density C and loss factor $\tan(\delta)$ as a function of the small-signal measurement frequency at zero bias voltage. b) Average discharged energy density $\langle w_d \rangle$ and efficiency η as a function of the number of charging/discharging cycles at 10.6 V. c) $\langle w_d \rangle$ as a function of the pulsed measurement frequency at 12.7 V. d) $\langle w_d \rangle$ and η as a function of temperature at 14.4 V. Adapted from Ref. [242].

The average discharged energy storage density $\langle w_d \rangle$ and efficiency η as a function of the number of charging/discharging cycles with an amplitude of 10.6 V are shown in Fig. 6.34b). Until 10^6 cycles, both $\langle w_d \rangle$ and η are stable at around 38 J cm^{-3} and 99 %, respectively. Only for more than 10^6 until 10^8 cycles, the stored energy decreases to 31 J cm^{-3}, while the efficiency stays constant. This effect might be related to pinning of ferroelectric domains with prolonged cycling, which is known as the fatigue effect [264]. In this case, the pinned domains cannot contribute to the NC anymore which reduces the overall stored energy. A different reason for the reduced storage might be successive charge injection to the ferroelectric/dielectric interface, which locally leads to switched domains with positive polarization, which have a positive capacitance when a positive voltage is applied.

Lastly, the robust temperature stability of $\langle w_d \rangle$ and η are shown in Fig. 6.34d) for an amplitude of 14.4 V. Overall, these initial reliability investigations are promising especially since these are the first experimental demonstrations of such NC supercapacitor devices, which have not been optimized for reliability so far.

For a further improvement of the energy storage density, dielectric materials with higher breakdown field strength are favorable. In addition, changing the ferroelectric and dielectric layer thicknesses, both the capacitance matching and overall storage density can be improved, since most of the energy is stored in the dielectric layer, as was shown in Fig. 6.31. Therefore, investigating the energy storage properties of the TiN/HZO/Al$_2$O$_3$/TiN samples discussed in Section 6.1 with different HZO and Al$_2$O$_3$ thicknesses is of interest.

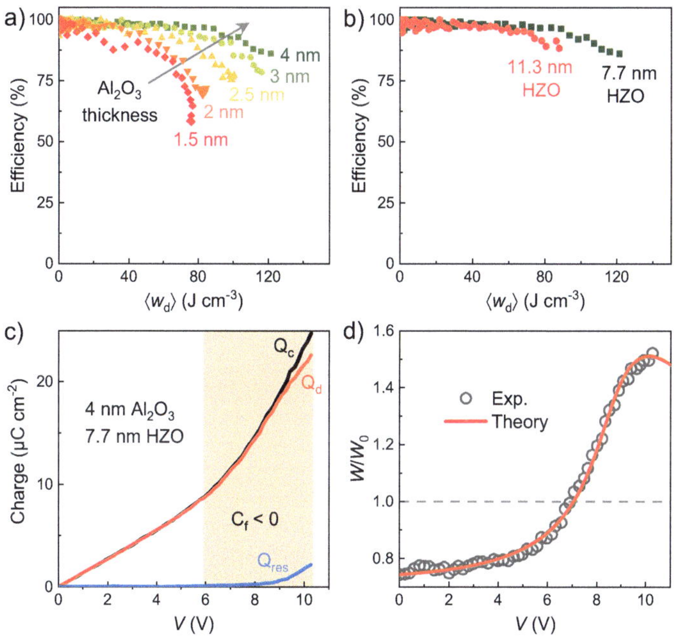

Figure 6.35: HZO/Al$_2$O$_3$-based NC supercapacitors with different layer thicknesses. a) Efficiency η as a function of the average discharged energy density $\langle w_d \rangle$ for 7.7 nm HZO and different Al$_2$O$_3$ layer thicknesses. b) η as a function of $\langle w_d \rangle$ for 4 nm Al$_2$O$_3$ and two different HZO thicknesses. c) Measured charges as a function of the applied voltage during charging (Q_c) and discharging (Q_d) for the 4 nm Al$_2$O$_3$/7.7 nm HZO sample. The difference is defined as $Q_{res} = Q_c - Q_d$. d) Comparison of experimental and theoretical energy storage enhancement W/W_0 as a function of voltage for the data in c). Adapted from Ref. [242].

Fig. 6.35a,b) show the efficiency vs. storage density curves for different Al$_2$O$_3$ and HZO thicknesses, respectively. Two major trends are obvious from these results. First, reducing the dielectric layer thickness strongly reduces the efficiency due to charge injection enabled by electron tunneling. If these charges are further trapped at the ferroelectric/dielectric interface, hysteretic ferroelectric switching can be induced leading to further reduced efficiency and storage density with cycling (negative reset pulses might be necessary). The second trend is the increase of the overall stored energy density with increasing the t_d/t_f ratio: Both increasing the Al_2O_3 thickness and decreasing the HZO thickness improves the maximum $\langle w_a \rangle$. The reason for this is that the average energy density is defined as the total energy W normalized to the whole capacitor volume $A(t_f + t_d)$ as

$$\langle w \rangle = \frac{W}{A(t_f + t_d)} = \frac{W_{FE} + W_{DE}}{A(t_f + t_d)} = \frac{\langle w_{FE} \rangle t_f + \langle w_{DE} \rangle t_d}{t_f + t_d}, \tag{6.40}$$

where W_{FE} ($\langle w_{FE} \rangle$) and W_{DE} ($\langle w_{DE} \rangle$) is the stored energy (density) in the ferroelectric and dielectric layer, respectively. Therefore, the overall average energy density $\langle w \rangle$ can be very different from the local energy densities $\langle w_{FE} \rangle$ and $\langle w_{DE} \rangle$. As was shown previously that for non-ideal capacitance matching, the stored energy in the dielectric W_{DE} is generally much larger than W_{FE}, which is why a reasonable approximation can be made as

$$\langle w \rangle \approx \frac{W_{DE}}{A(t_f + t_d)} = \frac{t_d}{t_f + t_d} \langle w_{DE} \rangle. \tag{6.41}$$

Eq. 6.41 shows that the overall calculated energy density $\langle w \rangle$ is smaller than $\langle w_{DE} \rangle$ roughly by a factor of $(1 + t_f/t_d)$. Therefore, to improve the overall energy density of the NC supercapacitor, the ratio t_f/t_d should be reduced. On the other hand, ideal capacitance matching demands $aC = -1$, and since $a = 2\alpha t_f/A$ and $C = \varepsilon_0 \varepsilon_d A/t_d$, one obtains

$$\frac{t_f}{t_d} = -\frac{1}{2\alpha\varepsilon_0\varepsilon_d}. \tag{6.42}$$

This means that generally, to reduce the ratio t_f/t_d while preserving ideal capacitance matching, dielectric materials with a higher relative permittivity and ferroelectrics with a larger absolute value of α are favorable. Nevertheless, one has to keep in mind that dielectrics with a higher permittivity typically have a reduced electronic band gap and therefore, reduced breakdown field strength, which also has to be considered when optimizing the material stack for an NC supercapacitor. The best overall energy storage density of 121 J cm^{-3} was achieved for the stack with 4 nm Al_2O_3 and 7.7 nm HZO shown in Fig. 6.35c) and d). The energy storage enhancement W/W_0 is roughly 1.5 at the maximum applied voltage, corresponding to a good capacitance matching of $aC \approx -0.58$, which is better than the one for the HZO/Ta_2O_5 stack in Fig. 6.32 of $aC \approx -0.16$. From the approximation in Eq. 6.41, the local maximum energy density in the Ta_2O_5 and Al_2O_3 layers can be estimated as 200 J cm^{-3} and 360 J cm^{-3}, respectively. This indicates that further improvement of the overall energy density should be feasible by optimizing the capacitance matching in conjunction with a reduction of the t_f/t_d ratio.

6.3.3 Comparison to literature

Most state-of-the-art solid-state electrostatic supercapacitors are based either on antiferroelectric or relaxor-like ferroelectric materials. To make a fair comparison to the best previous results, only non-epitaxial films are considered, since these can in principle be integrated on various 3D-structured substrates, which is a significant advantage compared to epitaxial films [265, 266, 267, 268, 269]. Fig. 6.36 compares the maximum average discharged energy densities for non-epitaxial electrostatic supercapacitors reported in literature [92, 220, 262, 270, 271, 272, 273, 274, 275, 276].

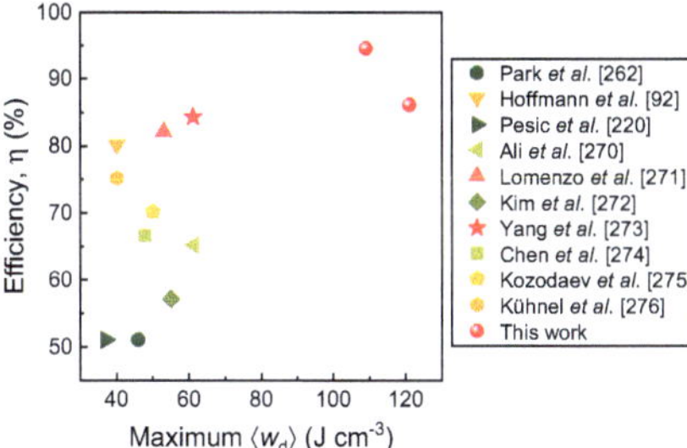

Figure 6.36: Comparison of maximum discharged energy densities $\langle w_d \rangle$ and corresponding efficiencies for nonepitaxial solid-state supercapacitors. Adapted from Ref. [242].

As can be seen, most results report a maximum energy density of 40-60 J cm^{-3}, with an efficiency below 85 %. At similar energy densities, the NC supercapacitors presented here have much higher efficiencies above 97 %. The maximum energy densities obtained for the NC supercapacitors are roughly twice as high compared to previous results, while at the same time achieving higher efficiencies. By using NC supercapacitors on 3D-structured substrates with high aspect ratio, even larger energy densities per projected substrate area can be envisioned and have recently been demonstrated in deep-trench capacitor structures [276].

6.4 Summary

Hysteresis-free NC was experimentally demonstrated under fast pulsed voltage operation in HZO/Al$_2$O$_3$ and HZO/Ta$_2$O$_5$ ferroelectric/dielectric capacitors. The use of voltage pulses in the range of 500 ns was necessary to impede charge injection to the ferroelectric/dielectric interface and breakdown of the dielectric layer. Hysteresis-free NC was observed down to pulse widths of 100 ns. Due to negative fixed charges at the ferroelectric/dielectric interface, the polarization of the HZO was stabilized pointing towards the ferroelectric/dielectric interface initially. The dielectric layer had to be of sufficient thickness (more than 3 nm) to prevent the screening of polarization charge by charge carrier injection and thus hysteretic switching. By calculating the field in the ferroelectric E_f and the polarization P, an 'S'-shaped P-E_f curve could be extracted from these electrical measurements in excellent agreement with Landau theory. Integration of the 'S'-shaped P-E_f curve enabled the calculation of the free energy-polarization landscape in the ferroelectric layer. The extracted energy barrier and coercive field were much lower than expected from first principles calculations and in accordance with standard electrical characterization methods on MFM capacitors. Therefore, it is still not clear if this hysteresis-free NC effect in ferroelectric HZO is of intrinsic or extrinsic nature. Further experiments and a better theoretical understanding of the ferroelectric switching process in HfO_2 based ferroelectric are needed. Dynamic simulations of the transient capacitor charging and discharging suggest that this NC effect is hysteresis-free even in the time-domain, when considering leakage currents through the capacitor and the correct parasitic series resistance.

Furthermore, it was proposed that similar MFIM capacitor structures with fixed interface charges are promising for energy storage applications in electrostatic supercapacitors. This NC supercapacitor concept has three main advantages over previous electrostatic supercapacitor concepts: 1. By engineering the charge-voltage nonlinearity, the stored energy can be increased by a factor of two compared to a regular capacitor even at identical voltage. 2. The breakdown voltage and leakage currents of the capacitor are improved by the addition of the NC layer, thus enabling much higher energy storage densities. 3. Since there is no hysteresis, the theoretical efficiency is 100 %. This concept was experimentally demonstrated by using HZO/Al_2O_3 and HZO/Ta_2O_5 ferroelectric/dielectric capacitors with different layer thicknesses. Since only polycrystalline and amorphous materials were used, integration on 3D substrates is in principle possible. Maximum discharged energy densities above $100\,J\,cm^{-3}$ were achieved at efficiencies higher than 95 %. Stable operation was shown in a broad frequency and temperature range (up to 150 °C) as well as for up to 10^8 charging/discharging cycles. Lastly, it was shown that increasing the ratio of dielectric to ferroelectric thickness can improve both maximum energy density as well as efficiency, since most of the energy is stored in the dielectric layer.

7 Conclusions and outlook

NC is an intriguing property of ferroelectric materials, which is deeply connected to the basic physics of ferroelectricity. Additionally, ferroelectric NC is promising for applications as it could be used to overcome the fundamental limitations of energy efficiency in electronics. However, the subtle differences in many experimental reports of ferroelectric NC, e.g. the role of domain formation and the observation of hysteresis, have not been well understood. Furthermore, with the relatively recent discovery of ferroelectricity in HfO_2 and ZrO_2 based thin films, research on NC in this new class of ferroelectrics is of significant practical interest for applications in nanoelectronic devices, where they might be used to overcome the Boltzmann limit of the subthreshold swing of $60 \, mV \, decade^{-1}$ at room temperature.

In the first part of this thesis (Chapter 2), the basic characteristics of ferroelectricity and NC were reviewed, concluding that one can generally distinguish four different types of NC phenomena in ferroelectrics: NC can either be stabilized (without a polarization hysteresis) or transient (with hysteresis) and its origin can either be intrinsic (from the thermodynamic polarization instability) or extrinsic (from the motion of domain walls). Furthermore, it was found that so far, HfO_2 and ZrO_2 based experimental NCFETs either exhibit no hysteresis and subthreshold swings close to the $60 \, mV \, decade^{-1}$ limit, or large hysteresis and subthreshold swings significantly lower than $60 \, mV \, decade^{-1}$. Further research into NC device design and the development of more accurate NC models for HfO_2 and ZrO_2 based ferroelectrics is therefore needed to improve practical NCFET devices.

In Chapter 4, transient NC in different metal-ferroelectric-metal capacitors was experimentally investigated. In polycrystalline HfO_2 based ferroelectric capacitors, it was found that transient NC could be observed when connecting a small resistance ($\leq 1 \, k\Omega$) in series and applying large voltage pulses. A decrease of the voltage across the ferroelectric during switching was interpreted as a transient NC effect. This behavior could be modeled with a MGLK approach, in which each ferroelectric grain switches individually according to a simplified Landau model. Furthermore, transient NC was observed in epitaxial PZT based capacitors showing extrinsic NC behavior caused by the nucleation and growth of domains in accordance with Kolmogorov-Avrami-Ishibashi switching dynamics. Similar NC voltage transients could be simulated by a numerical time-dependent Ginzburg-Landau model. By analyzing experiments with different series resistors, voltage amplitudes and capacitor areas, it was found that the time during which transient NC was observed is mainly limited by the speed of lateral domain wall motion and ends with domain coalescence. In the last part of Chapter 4, similar epitaxial PZT capacitors were connected in series with a dielectric capacitor, which enabled a direct measurement of the differential voltage amplification due to transient NC. All of these transient NC effects occur due to a partially unscreened ferroelectric polarization during switching. Overall, it was concluded that the large hysteresis observed in these transient NC experiments stems from the use of metal-ferroelectric-metal structures, which are undesirable for NC stabilization due to the small charge screening length in the metal electrodes. The fast screening of the ferroelectric polarization during switching in these structures can only be slowed down but not be inhibited.

In Chapter 5, the reason for this was further investigated when comparing the theory of intrinsic NC stabilization in structures with and without internal metal layer between the ferroelectric and a dielectric layer. In the structure with internal metal, due to the formation of anti-parallel domains, NC is fundamentally destabilized as a result of the electrostatic boundary conditions. Therefore, using an internal metal prevents the stabilization of intrinsic NC and is undesirable for NC devices. In contrast, when using a metal-ferroelectric-insulator-metal structure, it was found that intrinsic NC could be stabilized in principle, however, at a much lower critical ferroelectric thickness than expected from single-domain Landau theory. This sets a limit for the maximum possible voltage amplification due to intrinsic NC. Furthermore, it was found that by reducing the lateral device dimensions, the increase of the domain wall energy can help to improve the stabilization of intrinsic NC.

In Chapter 6, different ferroelectric/dielectric capacitors were fabricated using ferroelectric HZO and Al_2O_3 or Ta_2O_5 dielectrics to stabilize NC. However, at zero applied voltage, no stabilized NC effects were observed, which was attributed to trapped electrons at the ferroelectric/dielectric interface, reducing the depolarization field and stabilizing the spontaneous polarization in the negative remanent state. By applying short $\sim$500 ns positive voltage pulses to the capacitors, hysteresis-free (i.e. stabilized) NC was temporarily observed at higher applied voltages. An 'S'-shaped polarization-electric field dependence was extracted for the ferroelectric layer, corresponding to a double-well free energy landscape in accordance with Landau theory. While the extracted energy barrier from these measurements was much lower than predicted by first principles calculations, the experimental results show several characteristics of an intrinsic NC effect. Further work is necessary to confirm this. The absence of hysteresis was observed for dielectric layers thicker than 3 nm and short enough voltage pulses, such that charge injection to the ferroelectric/dielectric interface could be impeded. Furthermore, short pulses were necessary to prevent the dielectric breakdown of the capacitors due to the extremely high electric fields in the dielectric layer. It was shown that such hysteresis-free NC effects in HZO could be observed even for 100 ns short voltage pulses. Dynamical simulations indicate that this NC effect is hysteresis-free even in the time domain for short voltage pulses.

Furthermore, a new type of electrostatic supercapacitor was proposed based on the observed NC effect in ferroelectric/dielectric capacitors with fixed interface charges. This NC supercapacitor allows for an energy storage enhancement compared to a regular dielectric capacitor even at identical applied voltage, which would be impossible without NC. Since most of the energy in the NC supercapacitor is stored in the dielectric layer, the ratio of dielectric to ferroelectric thickness should be increased to improve the energy storage density. However, the energy storage enhancement is largest when the positive and negative capacitances of the layers are matched. Large storage densities above 100 J cm^{-3} and efficiencies higher than 95 % were experimentally shown. Furthermore, general frequency, temperature and cycling stability was demonstrated.

While transient and stabilized NC was shown to exist in ferroelectrics and especially in the application-relevant HfO$_2$ and ZrO$_2$ based ferroelectrics, there are still open questions about the microscopic origin of this behavior. Depending on whether the observed NC effects in HfO$_2$ and ZrO$_2$ based thin films are of intrinsic or extrinsic nature, the potential for applications might drastically change. Therefore, the fundamental limits of improvements expected for HfO$_2$ and ZrO$_2$ based NC devices are still unclear. While the current limited success of applying these materials in NCFET devices might hint a fundamental reason, it could also be related to the practical problem of fabricating phase-pure ultra-thin ferroelectric HfO$_2$ films, which might be necessary for highly scaled devices. The variability of polycrystalline ferroelectric films in such nanoscale devices is another concern for future research. Furthermore, the theoretical understanding of these NC effects is closely related to the basic switching kinetics and micro-structure of HfO$_2$ and ZrO$_2$ based ferroelectrics, which are still not completely understood. Future work has to reconcile the observed hysteresis-free NC effects shown here with other reports indicating nucleation-limited switching dynamics. This question is closely connected to the domain structure inside single grains of HfO$_2$ and ZrO$_2$ based ferroelectrics, which is not well understood at the moment. Even the fundamental domain wall energy constants for HfO$_2$ and ZrO$_2$ based ferroelectrics are not yet known, but might be accessible to first principles calculations or experiments in high-quality epitaxially grown capacitors in the future.

Lastly, as shown in this thesis for the case of NC supercapacitors, applications of ferroelectric NC beyond low power transistors can be envisioned. A multitude of uses of ferroelectric NC in sensors, actuators, memory devices etc. might be promising, even if the original idea of highly scaled low power transistors ultimately might prove impractical. Focused efforts of material scientists, physicists and electrical engineers are necessary to overcome these obstacles and pave the way towards real-world applications of ferroelectric NC.

Bibliography

[1] R. Landauer. Can Capacitance Be Negative? *Collective Phenomena*, 2: 167–170, 1976.

[2] T. N. Theis and P. M. Solomon. It's Time to Reinvent the Transistor! *Science*, 327: 1600–1601, 2010.

[3] G. E. Moore. Cramming More Componentsonto Integrated Circuits. *Electronics*, 38: 114–117, 1965.

[4] S. Salahuddin and S. Datta. Use of Negative Capacitance to Provide Voltage Amplification for Low Power Nanoscale Devices. *Nano Letters*, 8: 405–410, 2008.

[5] A. I. Khan. On the Microscopic Origin of Negative Capacitance in Ferroelectric Materials: A Toy Model. In *IEEE International Electron Devices Meeting (IEDM)*, pages 205–208. San Francisco, CA, USA, 2018.

[6] A. Khan, D. Bhowmik, P. Yu, S. Joo Kim, X. Pan, R. Ramesh, and S. Salahuddin. Experimental Evidence of Ferroelectric Negative Capacitance in Nanoscale Heterostructures. *Applied Physics Letters*, 99: 113501, 2011.

[7] P. Zubko, J. C. Wojdeł, M. Hadjimichael, S. Fernandez-Pena, A. Sené, I. Luk'yanchuk, J.-M. Triscone, and J. Íñiguez. Negative Capacitance in Multidomain Ferroelectric Superlattices. *Nature*, 534: 524–528, 2016.

[8] T. S. Böscke, J. Müller, D. Bräuhaus, U. Schröder, and U. Böttger. Ferroelectricity in Hafnium Oxide Thin Films. *Applied Physics Letters*, 99: 102903, 2011.

[9] J. Müller, T. S. Böscke, U. Schröder, S. Mueller, D. Bräuhaus, U. Böttger, L. Frey, and T. Mikolajick. Ferroelectricity in Simple Binary ZrO_2 and HfO_2. *Nano Letters*, 12: 4318–4323, 2012.

[10] M. E. Lines and A. M. Glass. *Principles and Applications of Ferroelectrics and Related Materials*. Oxford University Press, 2001.

[11] K. M. Rabe, C. H. Ahn, and J.-M. Triscone, editors. *Physics of Ferroelectrics: A Modern Perspective*. Topics in Applied Physics. Springer-Verlag, Berlin Heidelberg, 2007.

[12] A. K. Tagantsev. The Role of the Background Dielectric Susceptibility in Uniaxial Ferroelectrics. *Ferroelectrics*, 69: 321–323, 1986.

[13] T. Hahn, U. Shmueli, and J. W. Arthur. *International Tables for Crystallography*. Dordrecht: Reidel, 1983.

[14] R. Whatmore. Ferroelectric Materials. In S. Kasap and P. Capper, editors, *Springer Handbook of Electronic and Photonic Materials*, Springer Handbooks. Springer International Publishing, Cham, 2017.

[15] A. von Hippel. Piezoelectricity, Ferroelectricity, and Crystal Structure. *Zeitschrift für Physik A Hadrons and nuclei*, 133: 158–173, 1952.

[16] C. Kittel. Theory of Antiferroelectric Crystals. *Physical Review*, 82: 729–732, 1951.

[17] V. Maisonneuve, V. B. Cajipe, A. Simon, R. Von Der Muhll, and J. Ravez. Ferrielectric Ordering in Lamellar $CuInP_2S_6$. *Physical Review B*, 56: 10860–10868, 1997.

[18] L. E. Cross. Relaxor Ferroelectrics. *Ferroelectrics*, 76: 241–267, 1987.

[19] L. D. Landau. On the Theory of Phase Transitions. *Zhurnal Eksperimentalnoi i Teoreticheskoi Fiziki*, 7: 19–32, 1937.

[20] V. L. Ginzburg. On the Dielectric Properties of Ferroelectric (Segnette-Electric) Crystals and Barium Titanate. *Zhurnal Eksperimentalnoi i Teoreticheskoi Fiziki*, 15: 739–749, 1945.

[21] A. F. Devonshire. XCVI. Theory of Barium Titanate. *The London, Edinburgh, and Dublin Philosophical Magazine and Journal of Science*, 40: 1040–1063, 1949.

[22] A. F. Devonshire. Theory of Ferroelectrics. *Advances in Physics*, 3: 85–130, 1954.

[23] C. Woo and Y. Zheng. Depolarization in Modeling Nano-Scale Ferroelectrics Using the Landau Free Energy Functional. *Applied Physics A*, 91: 59–63, 2008.

[24] M. H. Park, T. Mikolajick, U. Schroeder, and C. S. Hwang. Broad Phase Transition of Fluorite-Structured Ferroelectrics for Large Electrocaloric Effect. *physica status solidi (RRL) – Rapid Research Letters*, 13: 1900177, 2019.

[25] K. Y. Yun, D. Ricinschi, T. Kanashima, M. Noda, and M. Okuyama. Giant Ferroelectric Polarization Beyond 150 $\mu C/cm^2$ in $BiFeO_3$ Thin Film. *Japanese Journal of Applied Physics*, 43: L647, 2004.

[26] I. P. Batra and B. D. Silverman. Thermodynamic Stability of Thin Ferroelectric Films. *Solid State Communications*, 11: 291–294, 1972.

[27] R. R. Mehta, B. D. Silverman, and J. T. Jacobs. Depolarization Fields in Thin Ferroelectric Films. *Journal of Applied Physics*, 44: 3379–3385, 1973.

[28] M. Dawber, P. Chandra, P. B. Littlewood, and J. F. Scott. Depolarization Corrections to the Coercive Field in Thin-Film Ferroelectrics. *Journal of Physics: Condensed Matter*, 15: L393–L398,

2003.

[29] M. E. Drougard and R. Landauer. On the Dependence of the Switching Time of Barium Titanate Crystals on Their Thickness. *Journal of Applied Physics*, 30: 1663–1668, 1959.

[30] A. M. Bratkovsky and A. P. Levanyuk. Abrupt Appearance of the Domain Pattern and Fatigue of Thin Ferroelectric Films. *Physical Review Letters*, 84: 3177–3180, 2000.

[31] M. Stengel and N. A. Spaldin. Origin of the Dielectric Dead Layer in Nanoscale Capacitors. *Nature*, 443: 679–682, 2006.

[32] P. Wurfel, I. P. Batra, and J. T. Jacobs. Polarization Instability in Thin Ferroelectric Films. *Physical Review Letters*, 30: 1218–1221, 1973.

[33] M. H. Park, H. J. Kim, Y. J. Kim, T. Moon, K. D. Kim, Y. H. Lee, S. D. Hyun, and C. S. Hwang. Study on the Internal Field and Conduction Mechanism of Atomic Layer Deposited Ferroelectric $Hf_{0.5}Zr_{0.5}O_2$ Thin Films. *J. Mater. Chem. C*, 3: 6291–6300, 2015.

[34] C. Lichtensteiger, P. Zubko, M. Stengel, P. Aguado-Puente, J.-M. Triscone, P. Ghosez, and J. Junquera. Ferroelectricity in Ultrathin-Film Capacitors. In *Oxide Ultrathin Films*, pages 265–230. John Wiley & Sons, Ltd, 2012.

[35] M. Hoffmann and T. Mikolajick. Verborgene Energielandschaften. *Physik in unserer Zeit*, 51, 2020. Article in press.

[36] A. Tagantsev, L. E. Cross, and J. Fousek. *Domains in Ferroic Crystals and Thin Films*. Springer-Verlag, New York, 2010.

[37] P. Wurfel and I. P. Batra. Depolarization-Field-Induced Instability in Thin Ferroelectric Films—Experiment and Theory. *Physical Review B*, 8: 5126–5133, 1973.

[38] V. Nagarajan, A. Roytburd, A. Stanishevsky, S. Prasertchoung, T. Zhao, L. Chen, J. Melngailis, O. Auciello, and R. Ramesh. Dynamics of Ferroelastic Domains in Ferroelectric Thin Films. *Nature Materials*, 2: 43–47, 2003.

[39] A. K. Yadav, C. T. Nelson, S. L. Hsu, Z. Hong, J. D. Clarkson, C. M. Schlepütz, A. R. Damodaran, P. Shafer, E. Arenholz, L. R. Dedon, D. Chen, A. Vishwanath, A. M. Minor, L. Q. Chen, J. F. Scott, L. W. Martin, and R. Ramesh. Observation of Polar Vortices in Oxide Superlattices. *Nature*, 530: 198–201, 2016.

[40] S. Das, Y. L. Tang, Z. Hong, M. a. P. Gonçalves, M. R. McCarter, C. Klewe, K. X. Nguyen, F. Gómez-Ortiz, P. Shafer, E. Arenholz, V. A. Stoica, S.-L. Hsu, B. Wang, C. Ophus, J. F. Liu, C. T. Nelson, S. Saremi, B. Prasad, A. B. Mei, D. G. Schlom, J. Íñiguez, P. García-Fernández, D. A. Muller, L. Q. Chen, J. Junquera, L. W. Martin, and R. Ramesh. Observation of Room-

Temperature Polar Skyrmions. *Nature*, 568: 368–372, 2019.

[41] A. Kopal, P. Mokrý, J. Fousek, and T. Bahník. Displacements of 180° Domain Walls in Electroded Ferroelectric Single Crystals: The Effect of Surface Layers on Restoring Force. *Ferroelectrics*, 223: 127–134, 1999.

[42] A. M. Bratkovsky and A. P. Levanyuk. Very Large Dielectric Response of Thin Ferroelectric Films with the Dead Layers. *Physical Review B*, 63: 132103, 2001.

[43] C. Ang and Z. Yu. Dielectric Behavior of $PbZr_{0.52}Ti_{0.48}O_3$ Thin Films: Intrinsic and Extrinsic Dielectric Responses. *Applied Physics Letters*, 85: 3821–3823, 2004.

[44] T. Schenk, M. Hoffmann, M. Pešić, M. H. Park, C. Richter, U. Schroeder, and T. Mikolajick. Physical Approach to Ferroelectric Impedance Spectroscopy: The Rayleigh Element. *Physical Review Applied*, 10: 064004, 2018.

[45] D. C. Lupascu, I. Anusca, M. Etier, Y. Gao, G. Lackner, A. Nazrabi, M. Sanlialp, H. Trivedi, N. Ul-Haq, and J. Schröder. Semiconductor Effects in Ferroelectrics. In J. Schröder and D. C. Lupascu, editors, *Ferroic Functional Materials: Experiment, Modeling and Simulation*, CISM International Centre for Mechanical Sciences, pages 97–178. Springer International Publishing, Cham, 2018.

[46] Z. C. Yuan, S. Rizwan, M. Wong, K. Holland, S. Anderson, T. B. Hook, D. Kienle, S. Gadelrab, P. S. Gudem, and M. Vaidyanathan. Switching-Speed Limitations of Ferroelectric Negative-Capacitance FETs. *IEEE Transactions on Electron Devices*, 63: 4046–4052, 2016.

[47] K. Karda, C. Mouli, and M. A. Alam. Switching Dynamics and Hot Atom Damage in Landau Switches. *IEEE Electron Device Letters*, 37: 801–804, 2016.

[48] Y. Li, K. Yao, and G. S. Samudra. Effect of Ferroelectric Damping on Dynamic Characteristics of Negative Capacitance Ferroelectric MOSFET. *IEEE Transactions on Electron Devices*, 63: 3636–3641, 2016.

[49] K. Chatterjee, A. J. Rosner, and S. Salahuddin. Intrinsic Speed Limit of Negative Capacitance Transistors. *IEEE Electron Device Letters*, 38: 1328–1330, 2017.

[50] K.-T. Chen, Y.-C. Chou, G.-Y. Siang, H.-Y. Chen, C. Lo, C.-Y. Liao, S.-T. Chang, and M.-H. Lee. Evaluation of Sweep Modes for Switch Response on Ferroelectric Negative-Capacitance FETs. *Applied Physics Express*, 12: 071003, 2019.

[51] H. W. Park, J. Roh, Y. B. Lee, and C. S. Hwang. Modeling of Negative Capacitance in Ferroelectric Thin Films. *Advanced Materials*, 31: 1805266, 2019.

[52] T. Schenk, E. Yurchuk, S. Mueller, U. Schroeder, S. Starschich, U. Böttger, and T. Mikolajick.

About the Deformation of Ferroelectric Hystereses. *Applied Physics Reviews*, 1: 041103, 2014.

[53] S. Ducharme, V. M. Fridkin, A. V. Bune, S. P. Palto, L. M. Blinov, N. N. Petukhova, and S. G. Yudin. Intrinsic Ferroelectric Coercive Field. *Physical Review Letters*, 84: 175–178, 2000.

[54] G. Vizdrik, S. Ducharme, V. M. Fridkin, and S. G. Yudin. Kinetics of Ferroelectric Switching in Ultrathin Films. *Physical Review B*, 68: 094113, 2003.

[55] M. J. Highland, T. T. Fister, M.-I. Richard, D. D. Fong, P. H. Fuoss, C. Thompson, J. A. Eastman, S. K. Streiffer, and G. B. Stephenson. Polarization Switching without Domain Formation at the Intrinsic Coercive Field in Ultrathin Ferroelectric PbTiO$_3$. *Physical Review Letters*, 105: 167601, 2010.

[56] J. F. Scott. Switching of Ferroelectrics Without Domains. *Advanced Materials*, 22: 5315–5317, 2010.

[57] W. J. Merz. Domain Formation and Domain Wall Motions in Ferroelectric BaTiO$_3$ Single Crystals. *Physical Review*, 95: 690–698, 1954.

[58] H. Orihara, S. Hashimoto, and Y. Ishibashi. A Theory of *D-E* Hysteresis Loop Based on the Avrami Model. *Journal of the Physical Society of Japan*, 63: 1031–1035, 1994.

[59] X. Du and I.-W. Chen. Frequency Spectra of Fatigue of PZT and Other Ferroelectric Thin Films. *MRS Online Proceedings Library Archive*, 493: 311–316, 1997.

[60] A. K. Tagantsev, I. Stolichnov, N. Setter, J. S. Cross, and M. Tsukada. Non-Kolmogorov-Avrami Switching Kinetics in Ferroelectric Thin Films. *Physical Review B*, 66: 214109, 2002.

[61] J. Valasek. Piezo-Electric and Allied Phenomena in Rochelle Salt. *Physical Review*, 17: 475–481, 1921.

[62] L. E. Cross and R. E. Newnham. History of Ferroelectrics. *Ceramics and Civilization*, III: 289–305, 1987.

[63] J. Grindlay, D. Ter Haar, and B. Bleaney. On the Ferroelectric Behaviour of Potassium Dihydrogen Phosphate. *Proceedings of the Royal Society of London. Series A. Mathematical and Physical Sciences*, 250: 266–285, 1959.

[64] A. von Hippel, R. G. Breckenridge, F. G. Chesley, and L. Tisza. High Dielectric Constant Ceramics. *Industrial & Engineering Chemistry*, 38: 1097–1109, 1946.

[65] T. Furukawa, M. Date, E. Fukada, Y. Tajitsu, and A. Chiba. Ferroelectric Behavior in the Copolymer of Vinylidenefluoride and Trifluoroethylene. *Japanese Journal of Applied Physics*, 19: L109–L112, 1980.

[66] B. T. Matthias, C. E. Miller, and J. P. Remeika. Ferroelectricity of Glycine Sulfate. *Physical Review*, 104: 849–850, 1956.

[67] G. Shirane and A. Takeda. Phase Transitions in Solid Solutions of $PbZrO_3$ and $PbTiO_3$ (I) Small Concentrations of $PbTiO_3$. *Journal of the Physical Society of Japan*, 7: 5–11, 1952.

[68] G. Shirane, K. Suzuki, and A. Takeda. Phase Transitions in Solid Solutions of $PbZrO_3$ and $PbTiO_3$ (II) X-Ray Study. *Journal of the Physical Society of Japan*, 7: 12–18, 1952.

[69] E. Sawaguchi, H. Maniwa, and S. Hoshino. Antiferroelectric Structure of Lead Zirconate. *Physical Review*, 83: 1078–1078, 1951.

[70] S. A. Mabud. The Morphotropic Phase Boundary in PZT Solid Solutions. *Journal of Applied Crystallography*, 13: 211–216, 1980.

[71] C.-L. Jia, V. Nagarajan, J.-Q. He, L. Houben, T. Zhao, R. Ramesh, K. Urban, and R. Waser. Unit-Cell Scale Mapping of Ferroelectricity and Tetragonality in Epitaxial Ultrathin Ferroelectric Films. *Nature Materials*, 6: 64–69, 2007.

[72] C. Paz de Araujo, J. D. Cuchiaro, L. D. McMillan, M. C. Scott, and J. F. Scott. Fatigue-Free Ferroelectric Capacitors with Platinum Electrodes. *Nature*, 374: 627–629, 1995.

[73] J. F. Scott. Applications of Modern Ferroelectrics. *Science*, 315: 954–959, 2007.

[74] C. Paz de Araujo, R. Ramesh, G. W. Taylor, R. Ramesh, and G. W. Taylor. *Science and Technology of Integrated Ferroelectrics: Selected Papers from Eleven Years of the Proceedings of the International Symposium of Integrated Ferroelectronics*. CRC Press, 2001.

[75] T. S. Böscke, J. Müller, D. Bräuhaus, U. Schröder, and U. Böttger. Ferroelectricity in Hafnium Oxide: CMOS Compatible Ferroelectric Field Effect Transistors. In *IEEE International Electron Devices Meeting (IEDM)*, pages 547–550. Washington, DC, USA, 2011.

[76] A. Kingon. Memories Are Made of . . . . *Nature*, 401: 658–659, 1999.

[77] J. Müller, U. Schröder, T. S. Böscke, I. Müller, U. Böttger, L. Wilde, J. Sundqvist, M. Lemberger, P. Kücher, T. Mikolajick, and L. Frey. Ferroelectricity in Yttrium-Doped Hafnium Oxide. *Journal of Applied Physics*, 110: 114113, 2011.

[78] S. Mueller, S. R. Summerfelt, J. Muller, U. Schroeder, and T. Mikolajick. Ten-Nanometer Ferroelectric Si:HfO_2 Films for Next-Generation FRAM Capacitors. *IEEE Electron Device Letters*, 33: 1300–1302, 2012.

[79] S. Mueller, J. Mueller, A. Singh, S. Riedel, J. Sundqvist, U. Schroeder, and T. Mikolajick. Incipient Ferroelectricity in Al-Doped HfO_2 Thin Films. *Advanced Functional Materials*, 22:

2412–2417, 2012.

[80] T. Schenk, S. Mueller, U. Schroeder, R. Materlik, A. Kersch, M. Popovici, C. Adelmann, S. Van Elshocht, and T. Mikolajick. Strontium Doped Hafnium Oxide Thin Films: Wide Process Window for Ferroelectric Memories. In *Proceedings of the European Solid-State Device Research Conference (ESSDERC)*, pages 260–263. Bucharest, Romania, 2013.

[81] U. Schroeder, C. Richter, M. H. Park, T. Schenk, M. Pešić, M. Hoffmann, F. P. G. Fengler, D. Pohl, B. Rellinghaus, C. Zhou, C.-C. Chung, J. L. Jones, and T. Mikolajick. Lanthanum-Doped Hafnium Oxide: A Robust Ferroelectric Material. *Inorganic Chemistry*, 57: 2752–2765, 2018.

[82] J. Müller, T. S. Böscke, D. Bräuhaus, U. Schröder, U. Böttger, J. Sundqvist, P. Kücher, T. Mikolajick, and L. Frey. Ferroelectric $Zr_{0.5}Hf_{0.5}O_2$ Thin Films for Nonvolatile Memory Applications. *Applied Physics Letters*, 99: 112901, 2011.

[83] M. Pešić, S. Knebel, M. Hoffmann, C. Richter, T. Mikolajick, and U. Schroeder. How to Make DRAM Non-Volatile? Anti-Ferroelectrics: A New Paradigm for Universal Memories. In *IEEE International Electron Devices Meeting (IEDM)*, pages 298–301. San Francisco, CA, USA, 2016.

[84] P. Polakowski and J. Müller. Ferroelectricity in Undoped Hafnium Oxide. *Applied Physics Letters*, 106: 232905, 2015.

[85] K. D. Kim, M. H. Park, H. J. Kim, Y. J. Kim, T. Moon, Y. H. Lee, S. D. Hyun, T. Gwon, and C. S. Hwang. Ferroelectricity in Undoped-HfO_2 Thin Films Induced by Deposition Temperature Control during Atomic Layer Deposition. *Journal of Materials Chemistry C*, 4: 6864–6872, 2016.

[86] C. Richter, T. Schenk, M. H. Park, F. A. Tscharntke, E. D. Grimley, J. M. LeBeau, C. Zhou, C. M. Fancher, J. L. Jones, T. Mikolajick, and U. Schroeder. Si Doped Hafnium Oxide—A "Fragile" Ferroelectric System. *Advanced Electronic Materials*, 3: 1700131, 2017.

[87] A. Navrotsky. Thermochemical Insights into Refractory Ceramic Materials Based on Oxides with Large Tetravalent Cations. *Journal of Materials Chemistry*, 15: 1883–1890, 2005.

[88] T. S. Böscke, S. Govindarajan, P. D. Kirsch, P. Y. Hung, C. Krug, B. H. Lee, J. Heitmann, U. Schröder, G. Pant, B. E. Gnade, and W. H. Krautschneider. Stabilization of Higher-κ Tetragonal HfO_2 by SiO_2 Admixture Enabling Thermally Stable Metal-Insulator-Metal Capacitors. *Applied Physics Letters*, 91: 072902, 2007.

[89] E. H. Kisi, C. J. Howard, and R. J. Hill. Crystal Structure of Orthorhombic Zirconia in Partially Stabilized Zirconia. *Journal of the American Ceramic Society*, 72: 1757–1760, 1989.

[90] T. S. Böscke, S. Teichert, D. Bräuhaus, J. Müller, U. Schröder, U. Böttger, and T. Mikolajick.

Phase Transitions in Ferroelectric Silicon Doped Hafnium Oxide. *Applied Physics Letters*, 99: 112904, 2011.

[91] M. Hoffmann, U. Schroeder, T. Schenk, T. Shimizu, H. Funakubo, O. Sakata, D. Pohl, M. Drescher, C. Adelmann, R. Materlik, A. Kersch, and T. Mikolajick. Stabilizing the Ferroelectric Phase in Doped Hafnium Oxide. *Journal of Applied Physics*, 118: 072006, 2015.

[92] M. Hoffmann, U. Schroeder, C. Künneth, A. Kersch, S. Starschich, U. Böttger, and T. Mikolajick. Ferroelectric Phase Transitions in Nanoscale HfO_2 Films Enable Giant Pyroelectric Energy Conversion and Highly Efficient Supercapacitors. *Nano Energy*, 18: 154–164, 2015.

[93] M. H. Park, T. Schenk, M. Hoffmann, S. Knebel, J. Gärtner, T. Mikolajick, and U. Schroeder. Effect of Acceptor Doping on Phase Transitions of HfO_2 Thin Films for Energy-Related Applications. *Nano Energy*, 36: 381–389, 2017.

[94] M. H. Park, C.-C. Chung, T. Schenk, C. Richter, M. Hoffmann, S. Wirth, J. L. Jones, T. Mikolajick, and U. Schroeder. Origin of Temperature-Dependent Ferroelectricity in Si-Doped HfO_2. *Advanced Electronic Materials*, 4: 1700489, 2018.

[95] S. E. Reyes-Lillo, K. F. Garrity, and K. M. Rabe. Antiferroelectricity in Thin Film ZrO_2 from First Principles. *Physical Review B*, 90: 140103, 2014.

[96] M. H. Park and C. S. Hwang. Fluorite-Structure Antiferroelectrics. *Reports on Progress in Physics*, 82: 124502, 2019.

[97] M. H. Park, Y. H. Lee, H. J. Kim, Y. J. Kim, T. Moon, K. D. Kim, J. Müller, A. Kersch, U. Schroeder, T. Mikolajick, and C. S. Hwang. Ferroelectricity and Antiferroelectricity of Doped Thin HfO_2-Based Films. *Advanced Materials*, 27: 1811–1831, 2015.

[98] S. Clima, D. J. Wouters, C. Adelmann, T. Schenk, U. Schroeder, M. Jurczak, and G. Pourtois. Identification of the Ferroelectric Switching Process and Dopant-Dependent Switching Properties in Orthorhombic HfO_2: A First Principles Insight. *Applied Physics Letters*, 104: 092906, 2014.

[99] T. D. Huan, V. Sharma, G. A. Rossetti, and R. Ramprasad. Pathways towards Ferroelectricity in Hafnia. *Physical Review B*, 90: 064111, 2014.

[100] R. Materlik, C. Künneth, and A. Kersch. The Origin of Ferroelectricity in $Hf_{1-x}Zr_xO_2$ - A Computational Investigation and a Surface Energy Model. *Journal of Applied Physics*, 117: 134109, 2015.

[101] M. Pešić, C. Künneth, M. Hoffmann, H. Mulaosmanovic, S. Müller, E. T. Breyer, U. Schroeder, A. Kersch, T. Mikolajick, and S. Slesazeck. A Computational Study of Hafnia-Based Ferro-

electric Memories: From Ab Initio via Physical Modeling to Circuit Models of Ferroelectric Device. *Journal of Computational Electronics*, 16: 1236–1256, 2017.

[102] B. Max, M. Hoffmann, S. Slesazeck, and T. Mikolajick. Direct Correlation of Ferroelectric Properties and Memory Characteristics in Ferroelectric Tunnel Junctions. *IEEE Journal of the Electron Devices Society*, 7: 1175–1181, 2019.

[103] F. Mehmood, M. Hoffmann, P. D. Lomenzo, C. Richter, M. Materano, T. Mikolajick, and U. Schroeder. Bulk Depolarization Fields as a Major Contributor to the Ferroelectric Reliability Performance in Lanthanum Doped $Hf_{0.5}Zr_{0.5}O_2$ Capacitors. *Advanced Materials Interfaces*, 6: 1901180, 2019.

[104] P. D. Lomenzo, S. Slesazeck, M. Hoffmann, T. Mikolajick, U. Schroeder, B. Max, and T. Mikolajick. Ferroelectric $Hf_{1-x}Zr_xO_2$ Memories: Device Reliability and Depolarization Fields. In *Non-Volatile Memory Technology Symposium (NVMTS)*, pages 1–8. Durham, NC, USA, 2019.

[105] M. Hoffmann, T. Schenk, M. Pešić, U. Schroeder, and T. Mikolajick. Insights into Antiferroelectrics from First-Order Reversal Curves. *Applied Physics Letters*, 111: 182902, 2017.

[106] X. Liu, H. Fan, J. Shi, and Q. Li. Origin of Anomalous Giant Dielectric Performance in Novel Perovskite: $Bi_{0.5-x}La_xNa_{0.5-x}Li_xTi_{1-y}M_yO_3$ (M = Mg^{2+}, Ga^{3+}). *Scientific Reports*, 5: 1–11, 2015.

[107] J. McPherson, J. Kim, A. Shanware, H. Mogul, and J. Rodriguez. Proposed Universal Relationship between Dielectric Breakdown and Dielectric Constant. In *IEEE International Electron Devices Meeting (IEDM)*, pages 633–636. San Francisco, CA, USA, 2002.

[108] L. Verman. Negative Circuit Constants. *Proceedings of the Institute of Radio Engineers*, 19: 676–681, 1931.

[109] W.-K. Chen. *The Circuits and Filters Handbook (Five Volume Slipcase Set)*. CRC Press, 2018.

[110] U. K. Chettiar, A. V. Kildishev, T. A. Klar, and V. M. Shalaev. Negative Index Metamaterial Combining Magnetic Resonators with Metal Films. *Optics Express*, 14: 7872–7877, 2006.

[111] S. A. Maier. Electromagnetics of Metals. In S. A. Maier, editor, *Plasmonics: Fundamentals and Applications*, pages 5–19. Springer US, New York, NY, 2007.

[112] S. A. Maier. Surface Plasmon Polaritons at Metal/Insulator Interfaces. In S. A. Maier, editor, *Plasmonics: Fundamentals and Applications*, pages 21–37. Springer US, New York, NY, 2007.

[113] S. A. Maier. Excitation of Surface Plasmon Polaritons at Planar Interfaces. In S. A. Maier, editor, *Plasmonics: Fundamentals and Applications*, pages 39–52. Springer US, New York, NY, 2007.

[114] S. A. Maier. Metamaterials and Imaging with Surface Plasmon Polaritons. In S. A. Maier, editor, *Plasmonics: Fundamentals and Applications*, pages 193–200. Springer US, New York, NY, 2007.

[115] M. Ershov, H. C. Liu, L. Li, M. Buchanan, Z. R. Wasilewski, and A. K. Jonscher. Negative Capacitance Effect in Semiconductor Devices. *IEEE Transactions on Electron Devices*, 45: 2196–2206, 1998.

[116] X. Wu, E. S. Yang, and H. L. Evans. Negative Capacitance at Metal-semiconductor Interfaces. *Journal of Applied Physics*, 68: 2845–2848, 1990.

[117] A. G. U. Perera, W. Z. Shen, M. Ershov, H. C. Liu, M. Buchanan, and W. J. Schaff. Negative Capacitance of GaAs Homojunction Far-Infrared Detectors. *Applied Physics Letters*, 74: 3167–3169, 1999.

[118] J. Bisquert, G. Garcia-Belmonte, Á. Pitarch, and H. J. Bolink. Negative Capacitance Caused by Electron Injection through Interfacial States in Organic Light-Emitting Diodes. *Chemical Physics Letters*, 422: 184–191, 2006.

[119] I. Mora-Seró, J. Bisquert, F. Fabregat-Santiago, G. Garcia-Belmonte, G. Zoppi, K. Durose, Y. Proskuryakov, I. Oja, A. Belaidi, T. Dittrich, R. Tena-Zaera, A. Katty, C. Lévy-Clément, V. Barrioz, and S. J. C. Irvine. Implications of the Negative Capacitance Observed at Forward Bias in Nanocomposite and Polycrystalline Solar Cells. *Nano Letters*, 6: 640–650, 2006.

[120] A. K. Jonscher. The Physical Origin of Negative Capacitance. *Journal of the Chemical Society, Faraday Transactions 2: Molecular and Chemical Physics*, 82: 75–81, 1986.

[121] H. Kam, D. Lee, R. Howe, and T.-J. King. A New Nano-Electro-Mechanical Field Effect Transistor (NEMFET) Design for Low-Power Electronics. In *IEEE International Electron Devices Meeting (IEDM)*, pages 463–466. Washington, DC, USA, 2005.

[122] N. Abele, R. Fritschi, K. Boucart, F. Casset, P. Ancey, and A. Ionescu. Suspended-Gate MOSFET: Bringing New MEMS Functionality into Solid-State MOS Transistor. In *IEEE International Electron Devices Meeting (IEDM)*, pages 479–481. Washington, DC, USA, 2005.

[123] A. Jain and M. A. Alam. Prospects of Hysteresis-Free Abrupt Switching (0 mV/Decade) in Landau Switches. *IEEE Transactions on Electron Devices*, 60: 4269–4276, 2013.

[124] M. A. Alam, M. Si, and P. D. Ye. A Critical Review of Recent Progress on Negative Capacitance Field-Effect Transistors. *Applied Physics Letters*, 114: 090401, 2019.

[125] A. M. Bratkovsky and A. P. Levanyuk. Depolarizing Field and "Real" Hysteresis Loops in Nanometer-Scale Ferroelectric Films. *Applied Physics Letters*, 89: 253108, 2006.

[126] J. Íñiguez, P. Zubko, I. Luk'yanchuk, and A. Cano. Ferroelectric Negative Capacitance. *Nature Reviews Materials*, 4: 243–256, 2019.

[127] A. I. Khan, K. Chatterjee, B. Wang, S. Drapcho, L. You, C. Serrao, S. R. Bakaul, R. Ramesh, and S. Salahuddin. Negative Capacitance in a Ferroelectric Capacitor. *Nature Materials*, 14: 182–186, 2015.

[128] M. Hoffmann, F. P. G. Fengler, M. Herzig, T. Mittmann, B. Max, U. Schroeder, R. Negrea, P. Lucian, S. Slesazeck, and T. Mikolajick. Unveiling the Double-Well Energy Landscape in a Ferroelectric Layer. *Nature*, 565: 464–467, 2019.

[129] M. Hoffmann, P. V. Ravindran, and A. I. Khan. Why Do Ferroelectrics Exhibit Negative Capacitance? *Materials*, 12: 3743, 2019.

[130] J. C. Wong and S. Salahuddin. Negative Capacitance Transistors. *Proceedings of the IEEE*, 107: 49–62, 2019.

[131] I. Luk'yanchuk, A. Sené, and V. M. Vinokur. Electrodynamics of Ferroelectric Films with Negative Capacitance. *Physical Review B*, 98: 024107, 2018.

[132] C. Kittel. Theory of the Structure of Ferromagnetic Domains in Films and Small Particles. *Physical Review*, 70: 965–971, 1946.

[133] J. Y. Son, S. Song, J.-H. Lee, and H. M. Jang. Anomalous Domain Periodicity Observed in Ferroelectric PbTiO$_3$ Nanodots Having 180° Stripe Domains. *Scientific Reports*, 6: 1–10, 2016.

[134] V. A. Stephanovich, I. A. Luk'yanchuk, and M. G. Karkut. Domain-Enhanced Interlayer Coupling in Ferroelectric/Paraelectric Superlattices. *Physical Review Letters*, 94: 047601, 2005.

[135] M. Hoffmann, M. Pešić, S. Slesazeck, U. Schroeder, and T. Mikolajick. On the Stabilization of Ferroelectric Negative Capacitance in Nanoscale Devices. *Nanoscale*, 10: 10891–10899, 2018.

[136] J. Wong. *Negative Capacitance and Hyperdimensional Computing for Unconventional Low-Power Computing*. Ph.D. thesis, EECS Department, University of California, Berkeley, 2018.

[137] M. Hoffmann, M. Pešić, K. Chatterjee, A. I. Khan, S. Salahuddin, S. Slesazeck, U. Schroeder, and T. Mikolajick. Direct Observation of Negative Capacitance in Polycrystalline Ferroelectric HfO$_2$. *Advanced Functional Materials*, 26: 8643–8649, 2016.

[138] S.-C. Chang, U. E. Avci, D. E. Nikonov, S. Manipatruni, and I. A. Young. Physical Origin of Transient Negative Capacitance in a Ferroelectric Capacitor. *Physical Review Applied*, 9: 014010, 2018.

[139] A. K. Saha, S. Datta, and S. K. Gupta. "Negative Capacitance" in Resistor-Ferroelectric and

Ferroelectric-Dielectric Networks: Apparent or Intrinsic? *Journal of Applied Physics*, 123: 105102, 2018.

[140] A. K. Yadav, K. X. Nguyen, Z. Hong, P. García-Fernández, P. Aguado-Puente, C. T. Nelson, S. Das, B. Prasad, D. Kwon, S. Cheema, A. I. Khan, C. Hu, J. Íñiguez, J. Junquera, L.-Q. Chen, D. A. Muller, R. Ramesh, and S. Salahuddin. Spatially Resolved Steady-State Negative Capacitance. *Nature*, 565: 468–471, 2019.

[141] M. Kobayashi. A Perspective on Steep-Subthreshold-Slope Negative-Capacitance Field-Effect Transistor. *Applied Physics Express*, 11: 110101, 2018.

[142] U. Schroeder, C. S. Hwang, and H. Funakubo. *Ferroelectricity in Doped Hafnium Oxide: Materials, Properties and Devices*. Elsevier, 1st edition, 2019.

[143] W. Cao and K. Banerjee. Is Negative Capacitance FET a Steep-Slope Logic Switch? *Nature Communications*, 11: 196, 2020.

[144] E. Morifuji, T. Yoshida, M. Kanda, S. Matsuda, S. Yamada, and F. Matsuoka. Supply and Threshold-Voltage Trends for Scaled Logic and SRAM MOSFETs. *IEEE Transactions on Electron Devices*, 53: 1427–1432, 2006.

[145] R. Dennard, F. Gaensslen, H.-N. Yu, V. Rideout, E. Bassous, and A. LeBlanc. Design of Ion-Implanted MOSFET's with Very Small Physical Dimensions. *IEEE Journal of Solid-State Circuits*, 9: 256–268, 1974.

[146] S. M. Sze and K. K. Ng. MOSFETs. In *Physics of Semiconductor Devices*, pages 293–373. John Wiley & Sons, Ltd, 2006.

[147] V. V. Zhirnov and R. K. Cavin. Negative Capacitance to the Rescue? *Nature Nanotechnology*, 3: 77–78, 2008.

[148] R. K. Jana, G. L. Snider, and D. Jena. On the Possibility of Sub 60 mV/Decade Subthreshold Switching in Piezoelectric Gate Barrier Transistors. *physica status solidi c*, 10: 1469–1472, 2013.

[149] A. M. Ionescu and H. Riel. Tunnel Field-Effect Transistors as Energy-Efficient Electronic Switches. *Nature*, 479: 329–337, 2011.

[150] K. Gopalakrishnan, P. Griffin, and J. Plummer. I-MOS: A Novel Semiconductor Device with a Subthreshold Slope Lower than kT/q. In *IEEE International Electron Devices Meeting (IEDM)*, pages 289–292. San Francisco, CA, USA, 2002.

[151] A. Padilla, C. W. Yeung, C. Shin, C. Hu, and Tsu-Jae King Liu. Feedback FET: A Novel Transistor Exhibiting Steep Switching Behavior at Low Bias Voltages. In *IEEE International Electron Devices Meeting (IEDM)*, pages 1–4. San Francisco, CA, USA, 2008.

[152] J. Ida, T. Mori, Y. Kuramoto, T. Horii, T. Yoshida, K. Takeda, H. Kasai, M. Okihara, and Y. Arai. Super Steep Subthreshold Slope PN-Body Tied SOI FET with Ultra Low Drain Voltage down to 0.1V. In *IEEE International Electron Devices Meeting (IEDM)*, pages 624–627. Washington, DC, USA, 2015.

[153] S. Mathews, R. Ramesh, T. Venkatesan, and J. Benedetto. Ferroelectric Field Effect Transistor Based on Epitaxial Perovskite Heterostructures. *Science*, 276: 238–240, 1997.

[154] A. Aziz, E. T. Breyer, A. Chen, X. Chen, S. Datta, S. K. Gupta, M. Hoffmann, X. S. Hu, A. Ionescu, M. Jerry, T. Mikolajick, H. Mulaosmanovic, K. Ni, M. Niemier, I. O'Connor, A. Saha, S. Slesazeck, S. K. Thirumala, and X. Yin. Computing with Ferroelectric FETs: Devices, Models, Systems, and Applications. In *Design, Automation Test in Europe Conference Exhibition (DATE)*, pages 1289–1298. Dresden, Germany, 2018.

[155] Y.-C. Chiu, C.-H. Cheng, C.-Y. Chang, M.-H. Lee, H.-H. Hsu, and S.-S. Yen. Low Power 1T DRAM/NVM Versatile Memory Featuring Steep Sub-60-mV/Decade Operation, Fast 20-ns Speed, and Robust 85°C-Extrapolated 10^{16} Endurance. In *Symposium on VLSI Technology (VLSI Technology)*, pages T184–T185. Kyoto, Japan, 2015.

[156] E. Yurchuk, J. Müller, S. Müller, J. Paul, M. Pešić, R. van Bentum, U. Schroeder, and T. Mikolajick. Charge-Trapping Phenomena in HfO_2-Based FeFET-Type Nonvolatile Memories. *IEEE Transactions on Electron Devices*, 63: 3501–3507, 2016.

[157] B. Obradovic, T. Rakshit, R. Hatcher, J. A. Kittl, and M. S. Rodder. Modeling Transient Negative Capacitance in Steep-Slope FeFETs. *IEEE Transactions on Electron Devices*, 65: 5157–5164, 2018.

[158] A. Daus, C. Vogt, N. Münzenrieder, L. Petti, S. Knobelspies, G. Cantarella, M. Luisier, G. A. Salvatore, and G. Tröster. Charge Trapping Mechanism Leading to Sub-60-mV/Decade-Swing FETs. *IEEE Transactions on Electron Devices*, 64: 2789–2796, 2017.

[159] G. A. Salvatore, D. Bouvet, and A. M. Ionescu. Demonstration of Subthrehold Swing Smaller than 60mV/Decade in Fe-FET with P(VDF-TrFE)/SiO_2 Gate Stack. In *IEEE International Electron Devices Meeting (IEDM)*, pages 1–4. San Francisco, CA, USA, 2008.

[160] A. Rusu, G. A. Salvatore, D. Jiménez, and A. M. Ionescu. Metal-Ferroelectric-Metal-Oxide-Semiconductor Field Effect Transistor with Sub-60mV. In *IEEE International Electron Devices Meeting (IEDM)*, pages 395–398. San Francisco, CA, USA, 2010.

[161] D. Jimenez, E. Miranda, and A. Godoy. Analytic Model for the Surface Potential and Drain Current in Negative Capacitance Field-Effect Transistors. *IEEE Transactions on Electron Devices*, 57: 2405–2409, 2010.

[162] A. I. Khan, C. W. Yeung, C. Hu, and S. Salahuddin. Ferroelectric Negative Capacitance MOSFET: Capacitance Tuning & Antiferroelectric Operation. In *IEEE International Electron Devices Meeting (IEDM)*, pages 255–258. Washington, DC, USA, 2011.

[163] D. J. Frank, P. M. Solomon, C. Dubourdieu, M. M. Frank, V. Narayanan, and T. N. Theis. The Quantum Metal Ferroelectric Field-Effect Transistor. *IEEE Transactions on Electron Devices*, 61: 2145–2153, 2014.

[164] K. Karda, A. Jain, C. Mouli, and M. A. Alam. An Anti-Ferroelectric Gated Landau Transistor to Achieve Sub-60 mV/Dec Switching at Low Voltage and High Speed. *Applied Physics Letters*, 106: 163501, 2015.

[165] J. P. Duarte, S. Khandelwal, A. I. Khan, A. Sachid, Y. K. Lin, H. L. Chang, S. Salahuddin, and C. Hu. Compact Models of Negative-Capacitance FinFETs: Lumped and Distributed Charge Models. In *IEEE International Electron Devices Meeting (IEDM)*, pages 754–757. San Francisco, CA, USA, 2016.

[166] M. Kobayashi and T. Hiramoto. On Device Design for Steep-Slope Negative-Capacitance Field-Effect-Transistor Operating at Sub-0.2V Supply Voltage with Ferroelectric HfO_2 Thin Film. *AIP Advances*, 6: 025113, 2016.

[167] Y.-H. Liao, S.-T. Fan, and C. W. Liu. Modeling and Simulation of Negative Capacitance Gate on Ge FETs. *ECS Transactions*, 75: 461–467, 2016.

[168] C.-I. Lin, A. I. Khan, S. Salahuddin, and C. Hu. Effects of the Variation of Ferroelectric Properties on Negative Capacitance FET Characteristics. *IEEE Transactions on Electron Devices*, 63: 2197–2199, 2016.

[169] H. Ota, T. Ikegami, J. Hattori, K. Fukuda, S. Migita, and A. Toriumi. Fully Coupled 3-D Device Simulation of Negative Capacitance FinFETs for Sub 10 nm Integration. In *IEEE International Electron Devices Meeting (IEDM)*, pages 318–321. San Francisco, CA, USA, 2016.

[170] G. Pahwa, T. Dutta, A. Agarwal, S. Khandelwal, S. Salahuddin, C. Hu, and Y. S. Chauhan. Analysis and Compact Modeling of Negative Capacitance Transistor with High ON-Current and Negative Output Differential Resistance—Part I: Model Description. *IEEE Transactions on Electron Devices*, 63: 4981–4985, 2016.

[171] G. Pahwa, T. Dutta, A. Agarwal, S. Khandelwal, S. Salahuddin, C. Hu, and Y. S. Chauhan. Analysis and Compact Modeling of Negative Capacitance Transistor with High ON-Current and Negative Output Differential Resistance—Part II: Model Validation. *IEEE Transactions on Electron Devices*, 63: 4986–4992, 2016.

[172] F. Liu, Y. Zhou, Y. Wang, X. Liu, J. Wang, and H. Guo. Negative Capacitance Transistors with

Monolayer Black Phosphorus. *npj Quantum Materials*, 1: 16004, 2016.

[173] S. Khandelwal, J. P. Duarte, A. I. Khan, S. Salahuddin, and C. Hu. Impact of Parasitic Capacitance and Ferroelectric Parameters on Negative Capacitance FinFET Characteristics. *IEEE Electron Device Letters*, 38: 142–144, 2017.

[174] H. Ota, S. Migita, J. Hattori, K. Fukuda, and A. Toriumi. Structural Advantages of Silicon-on-Insulator FETs over FinFETs in Steep Subthreshold-Swing Operation in Ferroelectric-Gate FETs. *Japanese Journal of Applied Physics*, 56: 04CD10, 2017.

[175] A. Sharma and K. Roy. Design Space Exploration of Hysteresis-Free $HfZrO_x$-Based Negative Capacitance FETs. *IEEE Electron Device Letters*, 38: 1165–1167, 2017.

[176] S. Pentapati, R. Perumal, S. Khandelwal, M. Hoffmann, S. K. Lim, and A. I. Khan. Cross-Domain Optimization of Ferroelectric Parameters for Negative Capacitance Transistors—Part I: Constant Supply Voltage. *IEEE Transactions on Electron Devices*, 67: 365–370, 2020.

[177] M.-H. Lee, J. C. Lin, Y. T. Wei, C. W. Chen, W. H. Tu, H. K. Zhuang, and M. Tang. Ferroelectric Negative Capacitance Hetero-Tunnel Field-Effect-Transistors with Internal Voltage Amplification. In *IEEE International Electron Devices Meeting (IEDM)*, pages 104–107. Washington, DC, USA, 2013.

[178] S. Dasgupta, A. Rajashekhar, K. Majumdar, N. Agrawal, A. Razavieh, S. Trolier-Mckinstry, and S. Datta. Sub-kT/q Switching in Strong Inversion in $PbZr_{0.52}Ti_{0.48}O_3$ Gated Negative Capacitance FETs. *IEEE Journal on Exploratory Solid-State Computational Devices and Circuits*, 1: 43–48, 2015.

[179] M. H. Lee, Y.-T. Wei, K.-Y. Chu, J.-J. Huang, C.-W. Chen, C.-C. Cheng, M.-J. Chen, H.-Y. Lee, Y.-S. Chen, L.-H. Lee, and M.-J. Tsai. Steep Slope and Near Non-Hysteresis of FETs With Antiferroelectric-Like HfZrO for Low-Power Electronics. *IEEE Electron Device Letters*, 36: 294–296, 2015.

[180] M. H. Lee, P.-G. Chen, K.-Y. Chu, C.-C. Cheng, M.-J. Xie, S.-N. Liu, J.-W. Lee, S.-J. Huang, M.-H. Liao, M. Tang, K.-S. Li, and M.-C. Chen. Prospects for Ferroelectric HfZrOx FETs with Experimentally CET=0.98nm, SSfor=42mVdec, SSrev=28mVdec, Switch-OFF 0.2V, and Hysteresis-Free Strategies. In *IEEE International Electron Devices Meeting (IEDM)*, pages 616–619. Washington, DC, USA, 2015.

[181] A. I. Khan, K. Chatterjee, J. P. Duarte, Z. Lu, A. Sachid, S. Khandelwal, R. Ramesh, C. Hu, and S. Salahuddin. Negative Capacitance in Short-Channel FinFETs Externally Connected to an Epitaxial Ferroelectric Capacitor. *IEEE Electron Device Letters*, 37: 111–114, 2016.

[182] F. A. McGuire, Z. Cheng, K. Price, and A. D. Franklin. Sub-60 mV/Decade Switching in 2D

Negative Capacitance Field-Effect Transistors with Integrated Ferroelectric Polymer. *Applied Physics Letters*, 109: 093101, 2016.

[183] C. Zhou and Y. Chai. Ferroelectric-Gated Two-Dimensional-Material-Based Electron Devices. *Advanced Electronic Materials*, 3: 1600400, 2016.

[184] E. Ko, J. W. Lee, and C. Shin. Negative Capacitance FinFET with Sub-20-mV/Decade Subthreshold Slope and Minimal Hysteresis of 0.48 V. *IEEE Electron Device Letters*, 38: 418–421, 2017.

[185] J. Zhou, G. Han, Y. Peng, Y. Liu, J. Zhang, Q. Q. Sun, D. W. Zhang, and Y. Hao. Ferroelectric Negative Capacitance GeSn PFETs With Sub-20 mV/Decade Subthreshold Swing. *IEEE Electron Device Letters*, 38: 1157–1160, 2017.

[186] M. Si, C. Jiang, W. Chung, Y. Du, M. A. Alam, and P. D. Ye. Steep-Slope WSe_2 Negative Capacitance Field-Effect Transistor. *Nano Letters*, 18: 3682–3687, 2018.

[187] X. Liu, R. Liang, G. Gao, C. Pan, C. Jiang, Q. Xu, J. Luo, X. Zou, Z. Yang, L. Liao, and Z. L. Wang. MoS_2 Negative-Capacitance Field-Effect Transistors with Subthreshold Swing below the Physics Limit. *Advanced Materials*, 30: 1800932, 2018.

[188] M. H. Lee, K.-T. Chen, C.-Y. Liao, S.-S. Gu, G.-Y. Siang, Y.-C. Chou, H.-Y. Chen, J. Le, R.-C. Hong, Z.-Y. Wang, S.-Y. Chen, P.-G. Chen, M. Tang, Y.-D. Lin, H.-Y. Lee, K.-S. Li, and C. W. Liu. Extremely Steep Switch of Negative-Capacitance Nanosheet GAA-FETs and FinFETs. In *IEEE International Electron Devices Meeting (IEDM)*, pages 735–738. San Francisco, CA, USA, 2018.

[189] X. Wang, P. Yu, Z. Lei, C. Zhu, X. Cao, F. Liu, L. You, Q. Zeng, Y. Deng, C. Zhu, J. Zhou, Q. Fu, J. Wang, Y. Huang, and Z. Liu. Van Der Waals Negative Capacitance Transistors. *Nature Communications*, 10: 3037, 2019.

[190] K.-S. Li, P.-G. Chen, T.-Y. Lai, C.-H. Lin, C.-C. Cheng, C.-C. Chen, Y.-J. Wei, Y.-F. Hou, M.-H. Liao, M.-H. Lee, et al. Sub-60mV-Swing Negative-Capacitance FinFET without Hysteresis. In *IEEE International Electron Devices Meeting (IEDM)*, pages 620–623. Washington, DC, USA, 2015.

[191] J. Jo and C. Shin. Negative Capacitance Field Effect Transistor With Hysteresis-Free Sub-60-mV/Decade Switching. *IEEE Electron Device Letters*, 37: 245–248, 2016.

[192] M. H. Lee, S. T. Fan, C. H. Tang, P. G. Chen, Y. C. Chou, H. H. Chen, J. Y. Kuo, M. J. Xie, S. N. Liu, M. H. Liao, C. A. Jong, K. S. Li, M. C. Chen, and C. W. Liu. Physical Thickness 1.x nm Ferroelectric HfZrOx Negative Capacitance FETs. In *IEEE International Electron Devices Meeting (IEDM)*, pages 306–309. San Francisco, CA, USA, 2016.

[193] J. Zhou, G. Han, Q. Li, Y. Peng, X. Lu, C. Zhang, J. Zhang, Q. Q. Sun, D. W. Zhang, and Y. Hao. Ferroelectric $HfZrO_x$ Ge and GeSn PMOSFETs with Sub-60 mV/Decade Subthreshold Swing, Negligible Hysteresis, and Improved I_{DS}. In *IEEE International Electron Devices Meeting (IEDM)*, pages 310–313. San Francisco, CA, USA, 2016.

[194] A. Nourbakhsh, A. Zubair, S. Joglekar, M. Dresselhaus, and T. Palacios. Subthreshold Swing Improvement in MoS_2 Transistors by the Negative-Capacitance Effect in a Ferroelectric Al-Doped-HfO_2/HfO_2 Gate Dielectric Stack. *Nanoscale*, 9: 6122–6127, 2017.

[195] Z. Krivokapic, U. Rana, R. Galatage, A. Razavieh, A. Aziz, J. Liu, J. Shi, H. J. Kim, R. Sporer, C. Serrao, A. Busquet, P. Polakowski, J. Müller, W. Kleemeier, A. Jacob, D. Brown, A. Knorr, R. Carter, and S. Banna. 14nm Ferroelectric FinFET Technology with Steep Subthreshold Slope for Ultra Low Power Applications. In *IEEE International Electron Devices Meeting (IEDM)*, pages 357–360. San Francisco, CA, USA, 2017.

[196] D. Kwon, K. Chatterjee, A. J. Tan, A. K. Yadav, H. Zhou, A. B. Sachid, R. D. Reis, C. Hu, and S. Salahuddin. Improved Subthreshold Swing and Short Channel Effect in FDSOI n-Channel Negative Capacitance Field Effect Transistors. *IEEE Electron Device Letters*, 39: 300–303, 2018.

[197] M. Si, C.-J. Su, C. Jiang, N. J. Conrad, H. Zhou, K. D. Maize, G. Qiu, C.-T. Wu, A. Shakouri, M. A. Alam, and P. D. Ye. Steep-Slope Hysteresis-Free Negative Capacitance MoS_2 Transistors. *Nature Nanotechnology*, 13: 24–28, 2018.

[198] D. Kwon, S. Cheema, N. Shanker, K. Chatterjee, Y.-H. Liao, A. J. Tan, C. Hu, and S. Salahuddin. Negative Capacitance FET With 1.8-Nm-Thick Zr-Doped HfO_2 Oxide. *IEEE Electron Device Letters*, 40: 993–996, 2019.

[199] Y.-H. Liao, D. Kwon, Y.-K. Lin, A. J. Tan, C. Hu, and S. Salahuddin. Anomalously Beneficial Gate-Length Scaling Trend of Negative Capacitance Transistors. *IEEE Electron Device Letters*, 40: 1860–1863, 2019.

[200] S.-Y. Lee, H.-W. Chen, C.-H. Shen, P.-Y. Kuo, C.-C. Chung, Y.-E. Huang, H.-Y. Chen, and T.-S. Chao. Experimental Demonstration of Stacked Gate- All-Around Poly-Si Nanowires Negative Capacitance FETs With Internal Gate Featuring Seed Layer and Free of Post-Metal Annealing Process. *IEEE Electron Device Letters*, 40: 1708–1711, 2019.

[201] D. Kwon, S. Cheema, Y.-K. Lin, Y.-H. Liao, K. Chatterjee, A. J. Tan, C. Hu, and S. Salahuddin. Near Threshold Capacitance Matching in a Negative Capacitance FET With 1 nm Effective Oxide Thickness Gate Stack. *IEEE Electron Device Letters*, 41: 179–182, 2020.

[202] Q. Han, T. C. U. Tromm, M. Hoffmann, P. Aleksa, U. Schroeder, J. Schubert, S. Mantl, and Q.-T. Zhao. Subthreshold Behavior of Floating-Gate MOSFETs With Ferroelectric Capacitors. *IEEE Transactions on Electron Devices*, 65: 4641–4645, 2018.

[203] K. Seshan. *Handbook of Thin Film Deposition.* Elsevier, fourth edition, 2018.

[204] D. Mattox. *Handbook of Physical Vapor Deposition (PVD) Processing.* Elsevier, second edition, 2010.

[205] M. Leskelä and M. Ritala. Atomic Layer Deposition (ALD): From Precursors to Thin Film Structures. *Thin Solid Films*, 409: 138–146, 2002.

[206] S. M. George. Atomic Layer Deposition: An Overview. *Chemical Reviews*, 110: 111–131, 2010.

[207] X. Sang, E. D. Grimley, T. Schenk, U. Schroeder, and J. M. LeBeau. On the Structural Origins of Ferroelectricity in HfO_2 Thin Films. *Applied Physics Letters*, 106: 162905, 2015.

[208] S. Verhaverbeke and J. W. Parker. A Model for the Etching of Ti and TiN in SC-1 Solutions. *MRS Online Proceedings Library Archive*, 477: 447–458, 1997.

[209] M. Birkholz. Grazing Incidence Configurations. In *Thin Film Analysis by X-Ray Scattering*, pages 143–182. John Wiley & Sons, Ltd, 2006.

[210] D. B. Williams and C. B. Carter. *Transmission Electron Microscopy.* Springer-Verlag US, Boston, MA, second edition, 2009.

[211] Y. J. Kim, H. Yamada, T. Moon, Y. J. Kwon, C. H. An, H. J. Kim, K. D. Kim, Y. H. Lee, S. D. Hyun, M. H. Park, and C. S. Hwang. Time-Dependent Negative Capacitance Effects in Al_2O_3/$BaTiO_3$ Bilayers. *Nano Letters*, 16: 4375–4381, 2016.

[212] S. Mueller, C. Adelmann, A. Singh, S. Van Elshocht, U. Schroeder, and T. Mikolajick. Ferroelectricity in Gd-Doped HfO_2 Thin Films. *ECS Journal of Solid State Science and Technology*, 1: N123–N126, 2012.

[213] M. Hoffmann, T. Schenk, I. Kulemanov, C. Adelmann, M. Popovici, U. Schroeder, and T. Mikolajick. Low Temperature Compatible Hafnium Oxide Based Ferroelectrics. *Ferroelectrics*, 480: 16–23, 2015.

[214] D. Martin, J. Müller, T. Schenk, T. M. Arruda, A. Kumar, E. Strelcov, E. Yurchuk, S. Müller, D. Pohl, U. Schröder, S. V. Kalinin, and T. Mikolajick. Ferroelectricity in Si-Doped HfO_2 Revealed: A Binary Lead-Free Ferroelectric. *Advanced Materials*, 26: 8198–8202, 2014.

[215] H. J. Kim, M. H. Park, Y. J. Kim, Y. H. Lee, W. Jeon, T. Gwon, T. Moon, K. D. Kim, and C. S. Hwang. Grain Size Engineering for Ferroelectric $Hf_{0.5}Zr_{0.5}O_2$ Films by an Insertion of Al_2O_3 Interlayer. *Applied Physics Letters*, 105: 192903, 2014.

[216] M. Pešić, C. Künneth, M. Hoffmann, H. Mulaosmanovic, S. Müller, E. T. Breyer, U. Schroeder, A. Kersch, T. Mikolajick, and S. Slesazeck. A Computational Study of Hafnia-Based Ferro-

electric Memories: From Ab Initio via Physical Modeling to Circuit Models of Ferroelectric Device. *Journal of Computational Electronics*, 16: 1236–1256, 2017.

[217] H. Mulaosmanovic, J. Ocker, S. Müller, U. Schroeder, J. Müller, P. Polakowski, S. Flachowsky, R. van Bentum, T. Mikolajick, and S. Slesazeck. Switching Kinetics in Nanoscale Hafnium Oxide Based Ferroelectric Field-Effect Transistors. *ACS Applied Materials & Interfaces*, 9: 3792–3798, 2017.

[218] M. Hoffmann, A. I. Khan, C. Serrao, Z. Lu, S. Salahuddin, M. Pešić, S. Slesazeck, U. Schroeder, and T. Mikolajick. Ferroelectric Negative Capacitance Domain Dynamics. *Journal of Applied Physics*, 123: 184101, 2018.

[219] Y. J. Kim, H. W. Park, S. D. Hyun, H. J. Kim, K. D. Kim, Y. H. Lee, T. Moon, Y. B. Lee, M. H. Park, and C. S. Hwang. Voltage Drop in a Ferroelectric Single Layer Capacitor by Retarded Domain Nucleation. *Nano Letters*, 17: 7796–7802, 2017.

[220] M. Pešić, M. Hoffmann, C. Richter, T. Mikolajick, and U. Schroeder. Nonvolatile Random Access Memory and Energy Storage Based on Antiferroelectric Like Hysteresis in ZrO_2. *Advanced Functional Materials*, 26: 7486–7494, 2016.

[221] M. Pešić, T. Li, V. Di Lecce, M. Hoffmann, M. Materano, C. Richter, B. Max, S. Slesazeck, U. Schroeder, L. Larcher, and T. Mikolajick. Built-In Bias Generation in Anti-Ferroelectric Stacks: Methods and Device Applications. *IEEE Journal of the Electron Devices Society*, 6: 1019–1025, 2018.

[222] T. Tybell, P. Paruch, T. Giamarchi, and J.-M. Triscone. Domain Wall Creep in Epitaxial Ferroelectric $Pb(Zr_{0.2}Ti_{0.8})O_3$ Thin Films. *Physical Review Letters*, 89: 097601, 2002.

[223] S. Liu, I. Grinberg, and A. M. Rappe. Intrinsic Ferroelectric Switching from First Principles. *Nature*, 534: 360–363, 2016.

[224] Y.-H. Shin, I. Grinberg, I.-W. Chen, and A. M. Rappe. Nucleation and Growth Mechanism of Ferroelectric Domain-Wall Motion. *Nature*, 449: 881–884, 2007.

[225] A. I. Khan, M. Hoffmann, K. Chatterjee, Z. Lu, R. Xu, C. Serrao, S. Smith, L. W. Martin, C. Hu, R. Ramesh, and S. Salahuddin. Differential Voltage Amplification from Ferroelectric Negative Capacitance. *Applied Physics Letters*, 111: 253501, 2017.

[226] G. Pahwa, T. Dutta, A. Agarwal, and Y. S. Chauhan. Physical Insights on Negative Capacitance Transistors in Nonhysteresis and Hysteresis Regimes: MFMIS Versus MFIS Structures. *IEEE Transactions on Electron Devices*, 65: 867–873, 2018.

[227] M. Hoffmann, M. Pešić, S. Slesazeck, U. Schroeder, and T. Mikolajick. Modeling and De-

sign Considerations for Negative Capacitance Field-Effect Transistors. In *Joint International EUROSOI Workshop and International Conference on Ultimate Integration on Silicon (EUROSOI-ULIS)*, pages 1–4. Athens, Greece, 2017.

[228] M. Hoffmann, S. Slesazeck, and T. Mikolajick. Domain Formation in Ferroelectric Negative Capacitance Devices. In *Device Research Conference (DRC)*, pages 1–2. Santa Barbara, CA, USA, 2018.

[229] I. Luk'yanchuk, Y. Tikhonov, A. Sené, A. Razumnaya, and V. M. Vinokur. Harnessing Ferroelectric Domains for Negative Capacitance. *Communications Physics*, 2: 1–6, 2019.

[230] A. I. Khan, U. Radhakrishna, K. Chatterjee, S. Salahuddin, and D. A. Antoniadis. Negative Capacitance Behavior in a Leaky Ferroelectric. *IEEE Transactions on Electron Devices*, 63: 4416–4422, 2016.

[231] W. Kinase and H. Takahasi. On the 180°-Type Domain Wall of $BaTiO_3$ Crystal. *Journal of the Physical Society of Japan*, 12: 464–476, 1957.

[232] Y. Cao, J. Shen, C. A. Randall, and L. Q. Chen. Phase-Field Modeling of Switchable Diode-like Current-Voltage Characteristics in Ferroelectric $BaTiO_3$. *Applied Physics Letters*, 104: 182905, 2014.

[233] X. Lu, H. Li, and W. Cao. Landau Expansion Parameters for $BaTiO_3$. *Journal of Applied Physics*, 114: 224106, 2013.

[234] R. K. Behera, C.-W. Lee, D. Lee, A. N. Morozovska, S. B. Sinnott, A. Asthagiri, V. Gopalan, and S. R. Phillpot. Structure and Energetics of 180° Domain Walls in $PbTiO_3$ by Density Functional Theory. *Journal of Physics: Condensed Matter*, 23: 175902, 2011.

[235] A. N. Morozovska, E. A. Eliseev, Y. Li, S. V. Svechnikov, P. Maksymovych, V. Y. Shur, V. Gopalan, L.-Q. Chen, and S. V. Kalinin. Thermodynamics of Nanodomain Formation and Breakdown in Scanning Probe Microscopy: Landau-Ginzburg-Devonshire Approach. *Physical Review B*, 80: 214110, 2009.

[236] M. J. Haun, E. Furman, S. J. Jang, H. A. McKinstry, and L. E. Cross. Thermodynamic Theory of $PbTiO_3$. *Journal of Applied Physics*, 62: 3331–3338, 1987.

[237] F. M. Pontes, E. J. H. Lee, E. R. Leite, E. Longo, and J. A. Varela. High Dielectric Constant of $SrTiO_3$ Thin Films Prepared by Chemical Process. *Journal of Materials Science*, 35: 4783–4787, 2000.

[238] A. Cano and D. Jiménez. Multidomain Ferroelectricity as a Limiting Factor for Voltage Amplification in Ferroelectric Field-Effect Transistors. *Applied Physics Letters*, 97: 133509,

2010.

[239] W. Gao, A. Khan, X. Marti, C. Nelson, C. Serrao, J. Ravichandran, R. Ramesh, and S. Salahuddin. Room-Temperature Negative Capacitance in a Ferroelectric–Dielectric Superlattice Heterostructure. *Nano Letters*, 14: 5814–5819, 2014.

[240] D. J. R. Appleby, N. K. Ponon, K. S. K. Kwa, B. Zou, P. K. Petrov, T. Wang, N. M. Alford, and A. O'Neill. Experimental Observation of Negative Capacitance in Ferroelectrics at Room Temperature. *Nano Letters*, 14: 3864–3868, 2014.

[241] M. Hoffmann, B. Max, T. Mittmann, U. Schroeder, S. Slesazeck, and T. Mikolajick. Demonstration of High-Speed Hysteresis-Free Negative Capacitance in Ferroelectric $Hf_{0.5}Zr_{0.5}O_2$. In *IEEE International Electron Devices Meeting (IEDM)*, pages 727–730. San Francisco, CA, USA, 2018.

[242] M. Hoffmann, F. P. G. Fengler, B. Max, U. Schroeder, S. Slesazeck, and T. Mikolajick. Negative Capacitance for Electrostatic Supercapacitors. *Advanced Energy Materials*, 9: 1901154, 2019.

[243] V. V. Afanas'ev, A. Stesmans, and W. Tsai. Determination of Interface Energy Band Diagram between (100)Si and Mixed Al–Hf Oxides Using Internal Electron Photoemission. *Applied Physics Letters*, 82: 245–247, 2003.

[244] H. C. Lin, P. D. Ye, and G. D. Wilk. Leakage Current and Breakdown Electric-Field Studies on Ultrathin Atomic-Layer-Deposited Al_2O_3 on GaAs. *Applied Physics Letters*, 87: 182904, 2005.

[245] M. H. Park, H. J. Kim, Y. J. Kim, W. Lee, T. Moon, and C. S. Hwang. Evolution of Phases and Ferroelectric Properties of Thin $Hf_{0.5}Zr_{0.5}O_2$ Films According to the Thickness and Annealing Temperature. *Applied Physics Letters*, 102: 242905, 2013.

[246] D. Zhou, J. Xu, Q. Li, Y. Guan, F. Cao, X. Dong, J. Müller, T. Schenk, and U. Schröder. Wake-up Effects in Si-Doped Hafnium Oxide Ferroelectric Thin Films. *Applied Physics Letters*, 103: 192904, 2013.

[247] T. Schenk, M. Hoffmann, J. Ocker, M. Pešić, T. Mikolajick, and U. Schroeder. Complex Internal Bias Fields in Ferroelectric Hafnium Oxide. *ACS Applied Materials & Interfaces*, 7: 20224–20233, 2015.

[248] F. P. G. Fengler, M. Hoffmann, S. Slesazeck, T. Mikolajick, and U. Schroeder. On the Relationship between Field Cycling and Imprint in Ferroelectric $Hf_{0.5}Zr_{0.5}O_2$. *Journal of Applied Physics*, 123: 204101, 2018.

[249] M. H. Park, H. J. Kim, Y. J. Kim, Y. H. Lee, T. Moon, K. D. Kim, S. D. Hyun, and C. S. Hwang. Study on the Size Effect in $Hf_{0.5}Zr_{0.5}O_2$ Films Thinner than 8 Nm before and after Wake-up

Field Cycling. *Applied Physics Letters*, 107: 192907, 2015.

[250] Y. J. Kim, M. H. Park, Y. H. Lee, H. J. Kim, W. Jeon, T. Moon, K. Do Kim, D. S. Jeong, H. Yamada, and C. S. Hwang. Frustration of Negative Capacitance in Al_2O_3/$BaTiO_3$ Bilayer Structure. *Scientific Reports*, 6: 19039, 2016.

[251] F. Werner, B. Veith, D. Zielke, L. Kühnemund, C. Tegenkamp, M. Seibt, R. Brendel, and J. Schmidt. Electronic and Chemical Properties of the c-Si/Al_2O_3 Interface. *Journal of Applied Physics*, 109: 113701, 2011.

[252] Y. Zhu, H. Ning, Z. Yu, Q. Pan, C. Zhang, C. Luo, X. Tu, Y. You, P. Wang, X. Wu, Y. Shi, and X. Wang. Thickness-Dependent Asymmetric Potential Landscape and Polarization Relaxation in Ferroelectric $Hf_xZr_{1-x}O_2$ Thin Films through Interfacial Bound Charges. *Advanced Electronic Materials*, 5: 1900554, 2019.

[253] M. Si, X. Lyu, and P. D. Ye. Ferroelectric Polarization Switching of Hafnium Zirconium Oxide in a Ferroelectric/Dielectric Stack. *ACS Applied Electronic Materials*, 1: 745–751, 2019.

[254] K. D. Kim, Y. J. Kim, M. H. Park, H. W. Park, Y. J. Kwon, Y. B. Lee, H. J. Kim, T. Moon, Y. H. Lee, S. D. Hyun, B. S. Kim, and C. S. Hwang. Transient Negative Capacitance Effect in Atomic-Layer-Deposited Al_2O_3/$Hf_{0.3}Zr_{0.7}O_2$ Bilayer Thin Film. *Advanced Functional Materials*, 29: 1808228, 2019.

[255] D. Jang, D. Yakimets, G. Eneman, P. Schuddinck, M. G. Bardon, P. Raghavan, A. Spessot, D. Verkest, and A. Mocuta. Device Exploration of NanoSheet Transistors for Sub-7-nm Technology Node. *IEEE Transactions on Electron Devices*, 64: 2707–2713, 2017.

[256] M. Hoffmann, S. Slesazeck, and T. Mikolajick. Dynamic Modeling of Hysteresis-Free Negative Capacitance in Ferroelectric/Dielectric Stacks under Fast Pulsed Voltage Operation. In *Device Research Conference (DRC)*. Ann Arbor, MI, USA, 2019.

[257] A. Evans, V. Strezov, and T. J. Evans. Assessment of Utility Energy Storage Options for Increased Renewable Energy Penetration. *Renewable and Sustainable Energy Reviews*, 16: 4141–4147, 2012.

[258] P. F. Ribeiro, B. K. Johnson, M. L. Crow, A. Arsoy, and Y. Liu. Energy Storage Systems for Advanced Power Applications. *Proceedings of the IEEE*, 89: 1744–1756, 2001.

[259] A. González, E. Goikolea, J. A. Barrena, and R. Mysyk. Review on Supercapacitors: Technologies and Materials. *Renewable and Sustainable Energy Reviews*, 58: 1189–1206, 2016.

[260] F. Gao, X. Dong, C. Mao, W. Liu, H. Zhang, L. Yang, F. Cao, and G. Wang. Energy-Storage Properties of $0.89Bi_{0.5}Na_{0.5}TiO_3$–$0.06BaTiO_3$–$0.05K_{0.5}Na_{0.5}NbO_3$ Lead-Free Anti-Ferroelectric

Ceramics. *Journal of the American Ceramic Society*, 94: 4382–4386, 2011.

[261] K. Yao, S. Chen, M. Rahimabady, M. S. Mirshekarloo, S. Yu, F. E. H. Tay, T. Sritharan, and L. Lu. Nonlinear Dielectric Thin Films for High-Power Electric Storage with Energy Density Comparable with Electrochemical Supercapacitors. *IEEE Transactions on Ultrasonics, Ferroelectrics, and Frequency Control*, 58: 1968–1974, 2011.

[262] M. H. Park, H. J. Kim, Y. J. Kim, T. Moon, K. D. Kim, and C. S. Hwang. Thin $Hf_xZr_{1-x}O_2$ Films: A New Lead-Free System for Electrostatic Supercapacitors with Large Energy Storage Density and Robust Thermal Stability. *Advanced Energy Materials*, 4: 1400610, 2014.

[263] M. McMillen, A. M. Douglas, T. M. Correia, P. M. Weaver, M. G. Cain, and J. M. Gregg. Increasing Recoverable Energy Storage in Electroceramic Capacitors Using "Dead-Layer" Engineering. *Applied Physics Letters*, 101: 242909, 2012.

[264] M. Pešić, F. P. G. Fengler, L. Larcher, A. Padovani, T. Schenk, E. D. Grimley, X. Sang, J. M. LeBeau, S. Slesazeck, U. Schroeder, and T. Mikolajick. Physical Mechanisms behind the Field-Cycling Behavior of HfO_2-Based Ferroelectric Capacitors. *Advanced Functional Materials*, 26: 4601–4612, 2016.

[265] H. Cheng, J. Ouyang, Y.-X. Zhang, D. Ascienzo, Y. Li, Y.-Y. Zhao, and Y. Ren. Demonstration of Ultra-High Recyclable Energy Densities in Domain-Engineered Ferroelectric Films. *Nature Communications*, 8: 1999, 2017.

[266] C. Hou, W. Huang, W. Zhao, D. Zhang, Y. Yin, and X. Li. Ultrahigh Energy Density in $SrTiO_3$ Film Capacitors. *ACS Applied Materials & Interfaces*, 9: 20484–20490, 2017.

[267] A. A. Instan, S. P. Pavunny, M. K. Bhattarai, and R. S. Katiyar. Ultrahigh Capacitive Energy Storage in Highly Oriented $Ba(Zr_xTi_{1-x})O_3$ Thin Films Prepared by Pulsed Laser Deposition. *Applied Physics Letters*, 111: 142903, 2017.

[268] Z. Sun, C. Ma, M. Liu, J. Cui, L. Lu, J. Lu, X. Lou, L. Jin, H. Wang, and C.-L. Jia. Ultrahigh Energy Storage Performance of Lead-Free Oxide Multilayer Film Capacitors via Interface Engineering. *Advanced Materials*, 29: 1604427, 2017.

[269] H. Pan, J. Ma, J. Ma, Q. Zhang, X. Liu, B. Guan, L. Gu, X. Zhang, Y.-J. Zhang, L. Li, Y. Shen, Y.-H. Lin, and C.-W. Nan. Giant Energy Density and High Efficiency Achieved in Bismuth Ferrite-Based Film Capacitors via Domain Engineering. *Nature Communications*, 9: 1813, 2018.

[270] F. Ali, X. Liu, D. Zhou, X. Yang, J. Xu, T. Schenk, J. Müller, U. Schroeder, F. Cao, and X. Dong. Silicon-Doped Hafnium Oxide Anti-Ferroelectric Thin Films for Energy Storage. *Journal of Applied Physics*, 122: 144105, 2017.

[271] P. D. Lomenzo, C.-C. Chung, C. Zhou, J. L. Jones, and T. Nishida. Doped $Hf_{0.5}Zr_{0.5}O_2$ for High Efficiency Integrated Supercapacitors. *Applied Physics Letters*, 110: 232904, 2017.

[272] K. D. Kim, Y. H. Lee, T. Gwon, Y. J. Kim, H. J. Kim, T. Moon, S. D. Hyun, H. W. Park, M. H. Park, and C. S. Hwang. Scale-up and Optimization of HfO_2-ZrO_2 Solid Solution Thin Films for the Electrostatic Supercapacitors. *Nano Energy*, 39: 390–399, 2017.

[273] B. B. Yang, M. Y. Guo, D. P. Song, X. W. Tang, R. H. Wei, L. Hu, J. Yang, W. H. Song, J. M. Dai, X. J. Lou, X. B. Zhu, and Y. P. Sun. Energy Storage Properties in $BaTiO_3$-$Bi_{3.25}La_{0.75}Ti_3O_{12}$ Thin Films. *Applied Physics Letters*, 113: 183902, 2018.

[274] J. Chen, Z. Tang, B. Yang, and S. Zhao. High Energy Storage Performances in Lead-Free $BaBi_{3.9}Pr_{0.1}Ti_4O_{15}$ Relaxor Ferroelectric Films. *Applied Physics Letters*, 113: 153904, 2018.

[275] M. G. Kozodaev, A. G. Chernikova, R. R. Khakimov, M. H. Park, A. M. Markeev, and C. S. Hwang. La-Doped $Hf_{0.5}Zr_{0.5}O_2$ Thin Films for High-Efficiency Electrostatic Supercapacitors. *Applied Physics Letters*, 113: 123902, 2018.

[276] K. Kühnel, M. Czernohorsky, C. Mart, and W. Weinreich. High-Density Energy Storage in Si-Doped Hafnium Oxide Thin Films on Area-Enhanced Substrates. *Journal of Vacuum Science & Technology B*, 37: 021401, 2019.

List of symbols

Symbol	Description	Units
a	1st normalized Landau coefficient	F^{-1}
A	capacitor area	m^2
A_V	voltage amplification	1
α	1st Landau coefficient	$\mathrm{m\,F}^{-1}$
α_0	0th Landau coefficient	$\mathrm{m\,F}^{-1}\,\mathrm{K}^{-1}$
b	2nd normalized Landau coefficient	$\mathrm{F}^{-1}\,\mathrm{C}^{-2}$
β	2nd Landau coefficient	$\mathrm{m}^5\,\mathrm{F}^{-1}\,\mathrm{C}^{-2}$
c	3rd normalized Landau coefficient	$\mathrm{F}^{-1}\,\mathrm{C}^{-4}$
C	capacitance	F
C_0	capacitance constant	$\mathrm{F\,m}^{-2}$
c_A	area dependence fitting slope	$\mathrm{s\,m}^{-2}$
C_d	dielectric capacitance per unit area	$\mathrm{F\,m}^{-2}$
C_D	dielectric capacitance	F
C_f	ferroelectric capacitance per unit area	$\mathrm{F\,m}^{-2}$
C_F	ferroelectric capacitance	F
$C_{f,b}$	ferroelectric background capacitance per unit area	$\mathrm{F\,m}^{-2}$
$C_{F,b}$	ferroelectric background capacitance	F
C_{ins}	gate insulator capacitance	$\mathrm{F\,m}^{-2}$
C_p	total parasitic capacitance	F
C_s	semiconductor capacitance	$\mathrm{F\,m}^{-2}$
C_{sample}	parasitic capacitance of the sample	F
C_{self}	self-capacitance of a conductor	F
C_{setup}	parasitic capacitance of the experimental setup	F
χ	electric susceptibility	1
χ_b	ferroelectric background susceptibility	1
d	domain period	m
D	electric displacement field	$\mathrm{C\,m}^{-2}$
d_c	distance between the electrodes of a capacitor	m
D_d	electric displacement field in the dielectric layer	$\mathrm{C\,m}^{-2}$
d_{eq}	equilibrium domain period	m
D_n	multi-domain permittivity coefficients	1
δ	loss angle	$^\circ$
ΔQ	difference in electric charge	C
ΔQ_d	discharging difference between two voltage pulses	C
Δt	simulation time step	s
ΔT	internal bias corrected negative capacitance transient time	s
ΔT_n	negative capacitance transient time for negative applied voltage	s
ΔT_p	negative capacitance transient time for positive applied voltage	s

Symbol	Description	Units
ΔV_{max}	voltage amplitude difference between two voltage pulses	V
ΔV_n	negative capacitance voltage drop for negative applied voltage	V
ΔV_p	negative capacitance voltage drop for positive applied voltage	V
Δx	simulation grid spacing in x-direction	m
Δz	simulation grid spacing in z-direction	m
E	electric field	$\mathrm{V\,m^{-1}}$
E_a	activation field	$\mathrm{V\,m^{-1}}$
E_{bias}	internal bias field	$\mathrm{V\,m^{-1}}$
E_c	coercive field	$\mathrm{V\,m^{-1}}$
E_{c0}	intrinsic coercive field	$\mathrm{V\,m^{-1}}$
E_d	total electric field in the dielectric	$\mathrm{V\,m^{-1}}$
E_{dep}	depolarization field	$\mathrm{V\,m^{-1}}$
$E_{d,ext}$	external electric field in the dielectric	$\mathrm{V\,m^{-1}}$
$E_{d,int}$	internal electric field in the dielectric	$\mathrm{V\,m^{-1}}$
E_f	total electric field in the ferroelectric	$\mathrm{V\,m^{-1}}$
$E_{f,ext}$	external electric field in the ferroelectric	$\mathrm{V\,m^{-1}}$
$\langle E_{dep,h} \rangle$	average homogeneous depolarization field in the ferroelectric	$\mathrm{V\,m^{-1}}$
$\langle E_f \rangle$	average total electric field in the ferroelectric	$\mathrm{V\,m^{-1}}$
ϵ	local relative error of the iterative Poisson solver	1
ε_0	vacuum permittivity	$\mathrm{F\,m^{-1}}$
ε_b	relative ferroelectric background permittivity	1
$\varepsilon_{b,x}$	in-plane relative ferroelectric background permittivity	1
$\varepsilon_{b,z}$	out-of-plane relative ferroelectric background permittivity	1
ε_d	relative permittivity of the dielectric	1
ε_f	relative ferroelectric permittivity	1
η	charging/discharging efficiency	1
f	frequency	Hz
F	Helmholtz free energy density per unit area	$\mathrm{J\,m^{-2}}$
F_0	free energy integration constant	$\mathrm{J\,m^{-3}}$
F_a	ferroelectric anisotropy free energy density	$\mathrm{J\,m^{-3}}$
F_d	dielectric Helmholtz free energy density	$\mathrm{J\,m^{-3}}$
F_{DW}	total domain wall energy density	$\mathrm{J\,m^{-3}}$
F_f	ferroelectric Helmholtz free energy density	$\mathrm{J\,m^{-3}}$
$F_{single,DW}$	domain wall energy density per domain wall area	$\mathrm{J\,m^{-2}}$
$f(x)$	differentiable function of variable x	1
G	Gibb's free energy density	$\mathrm{J\,m^{-3}}$
γ	3rd Landau coefficient	$\mathrm{m^9\,F^{-1}\,C^{-4}}$
i	domain index	1
I	electric current	A
(i)	iteration step in the iterative Poisson solver	1

Symbol	Description	Units
I_{Cp}	parasitic capacitor current	A
I_d	drain current	A
I_F	ferroelectric current	A
I_L	leakage current	A
I_{L0}	leakage current fitting constant	A
I_{off}	off-state drain current	A
I_{on}	on-state drain current	A
I_R	current flowing through the resistor R	A
j	grain index	1
k	domain wall coupling constant	$\text{m}^3\,\text{F}^{-1}$
k_1	Fowler-Nordheim constant 1	$\text{A}\,\text{V}^{-2}$
k_2	Fowler-Nordheim constant 2	V
k_B	Boltzmann constant	$\text{J}\,\text{K}^{-1}$
l	index of adjacent domains	1
L	lateral capacitor dimension	m
L_{crit}	critical lateral capacitor dimension	m
λ	wavelength of Cu-Kα radiation	m
m	exponent in Merz's law	1
μ	mean value	1
μ_{DW}	domain wall mobility	$\text{m}^2\,\text{V}^{-1}\,\text{s}^{-1}$
n	natural number	1
N	number of ferroelectric grains	1
N_{DW}	number of domain walls	1
ω	angular frequency	$\text{rad}\,\text{s}^{-1}$
P	electric polarization	$\text{C}\,\text{m}^{-2}$
P_0	spontaneous ferroelectric polarization at zero electric field	$\text{C}\,\text{m}^{-2}$
P_1	spontaneous polarization of domain 1	$\text{C}\,\text{m}^{-2}$
P_2	spontaneous polarization of domain 2	$\text{C}\,\text{m}^{-2}$
P_b	ferroelectric background polarization	$\text{C}\,\text{m}^{-2}$
P_{diss}	power dissipation	$\text{J}\,\text{s}^{-1}$
P_{max}	maximum polarization	$\text{C}\,\text{m}^{-2}$
P_{min}	minimum polarization	$\text{C}\,\text{m}^{-2}$
P_{offset}	polarization offset	$\text{C}\,\text{m}^{-2}$
P_r	remanent polarization	$\text{C}\,\text{m}^{-2}$
P_S	spontaneous ferroelectric polarization	$\text{C}\,\text{m}^{-2}$
P_z	spontaneous polarization in z-direction	$\text{C}\,\text{m}^{-2}$
$\langle P_S \rangle$	average spontaneous polarization	$\text{C}\,\text{m}^{-2}$
φ	electrostatic potential	V
φ_1	electrostatic potential of conductor 1	V
φ_2	electrostatic potential of conductor 2	V

Symbol	Description	Units
φ_{zx}	discretized electrostatic potential at position (z,x)	V
ψ_s	semiconductor surface potential	V
q	elementary charge	C
Q	electric charge	C
Q_0	reference charge of a linear capacitor	C
Q_c	total charge during capacitor charging	C
Q_{Cp}	charge on the parasitic capacitor	C
Q_d	total charge during capacitor discharging	C
Q_F	ferroelectric charge	C
Q_{F0}	initial ferroelectric charge	C
Q_{res}	residual charge	C
Q_x	stored charge for maximum energy storage enhancement	C
r	atomic ratio	1
R	resistance	Ω
R^2	coefficient of determination	1
R_{epr}	equivalent parallel resistance	Ω
R_i	internal ferroelectric resistance	Ω
R_L	Leakage resistance of the ferroelectric	Ω
ρ	Landau-Khalatnikov damping constant	$\Omega\,\mathrm{m}$
ρ_{free}	free charge density per volume	$\mathrm{C\,m^{-3}}$
ρ_{max}	maximum Landau-Khalatnikov damping constant	$\Omega\,\mathrm{m}$
ρ_{min}	minimum Landau-Khalatnikov damping constant	$\Omega\,\mathrm{m}$
ρ_{vol}	charge density per volume	$\mathrm{C\,m^{-3}}$
S	subthreshold swing	$\mathrm{V\,decade^{-1}}$
s_0	Landau-Lifshitz-Kittel scaling constant	m
σ	standard deviation	1
σ_{IF}	interfacial charge density	$\mathrm{C\,m^{-2}}$
t	time	s
T	absolute temperature	K
T_0	Curie-Weiss temperature	K
T_1	limit-temperature of ferroelectricity	K
T_2	limit-temperature of field-induced ferroelectricity	K
T_C	Curie-temperature	K
t_d	dielectric layer thickness	m
t_f	ferroelectric layer thickness	m
T_f	voltage pulse fall time	s
$t_{f,dw}$	critical ferroelectric thickness for domain formation	m
$t_{f,max}$	critical ferroelectric thickness	m
t_{nmax}	start time of ferroelectric voltage drop during negative switching	s
t_{nmin}	end time of ferroelectric voltage drop during negative switching	s

Symbol	Description	Units
t_{pmax}	start time of ferroelectric voltage drop during positive switching	s
t_{pmin}	end time of ferroelectric voltage drop during positive switching	s
T_{pulse}	voltage pulse width	s
T_r	voltage pulse rise time	s
T_{step}	voltage time step	s
τ_0	intrinsic switching time constant	s
τ_A	area dependence fitting offset	s
τ_{sw}	characteristic switching time	s
2θ	detector angle in grazing-incidence X-ray diffraction	°
U	Gibb's free energy	J
V	voltage	V
v_{AC}	AC voltage signal	V
V_{AC}	small-signal voltage amplitude	V
V_{max}	maximum voltage amplitude	V
V_b	internal bias voltage	V
V_D	voltage across the dielectric layer	V
V_{dd}	power supply voltage	V
V_{ds}	drain-source voltage	V
V_{DUT}	voltage across the device under test	V
v_{DW}	domain wall velocity	m s^{-1}
V_F	voltage across the ferroelectric	V
V_{gs}	gate-source voltage	V
V_i	internal ferroelectric voltage	V
V_{L0}	leakage current voltage fitting constant	V
V_{nmax}	maximum ferroelectric voltage at onset of negative switching	V
V_{nmin}	minimum ferroelectric voltage during negative switching	V
V_{pmax}	maximum ferroelectric voltage at onset of positive switching	V
V_{pmin}	minimum ferroelectric voltage during positive switching	V
V_R	voltage drop across the resistance R	V
V_S	applied voltage signal	V
V_{step}	voltage step size	V
V_x	applied voltage for maximum energy storage enhancement	V
w	electrostatic energy density	J m^{-3}
W	electrostatic energy	J
W_0	reference electrostatic energy of a linear capacitor	J
W_c	electrostatic charging energy	J
W_d	released electrostatic energy during discharging	J
W_{DE}	stored energy in the dielectric	J
w_{dep}	depolarization energy density	J m^{-3}
w_{dw}	domain wall width	m

Symbol	Description	Units
W_{FE}	stored energy in the ferroelectric	J
W_{loss}	dissipated energy during charging and discharging	J
W_{NC}	electrostatic energy stored in the negative capacitance layer	J
W_{PC}	electrostatic energy stored in the positive capacitance layer	J
W_x	energy stored at maximum energy storage enhancement	J
$\langle w_c \rangle$	average electrostatic charging energy density	$\mathrm{J\,m^{-3}}$
$\langle w_d \rangle$	average discharged energy density	$\mathrm{J\,m^{-3}}$
$\langle w_{DE} \rangle$	average stored energy density in the dielectric	$\mathrm{J\,m^{-3}}$
$\langle w_{FE} \rangle$	average stored energy density in the ferroelectric	$\mathrm{J\,m^{-3}}$
$\langle w_{loss} \rangle$	average dissipated energy density	$\mathrm{J\,m^{-3}}$
x	spatial coordinate in the in-plane direction	m
X	simulation grid size in x-direction	1
y	spatial coordinate in the in-plane direction	m
z	spatial coordinate in the out-of-plane direction	m
Z	simulation grid size in z-direction	1

List of abbreviations

Abbreviation	Description
AC	alternating current
ALD	atomic layer deposition
CMOS	complementary metal-oxide-semiconductor
DC	direct current
DRAM	dynamic random access memory
DUT	device under test
EELS	electron energy loss spectroscopy
FeFET	ferroelectric field-effect transistor
GIXRD	grazing incidence X-ray diffraction
HZO	hafnium zirconium oxide
LGD	Landau-Ginzburg-Devonshire
MFIM	metal-ferroelectric-insulator-metal
MFM	metal-ferroelectric-metal
MFMIM	metal-ferroelectric-metal-insulator-metal
MGLK	multi-grain Landau Khalatnikov
MOSFET	metal-oxide-semiconductor field-effect transistor
NC	negative capacitance
NCFET	negative capacitance field-effect transistor
PLD	pulsed laser deposition
PZT	lead zirconate titanate
RT	room temperature
SAED	selected area electron diffraction
SC-1	standard clean one
TEM	transmission electron microscopy
TEMAHf	Tetrakis(ethylmethylamino)hafnium
TEMAZr	Tetrakis(ethylmethylamino)zirconium
TMA	Trimethylaluminum
XRD	X-ray diffraction

Acknowledgments

Here, I want to thank everyone who either directly or indirectly supported me during the completion of this thesis. First and foremost, I would like to thank Prof. Thomas Mikolajick for supervising this thesis, providing helpful advice as well as creating an environment with generous freedoms to pursue my own research interests. Secondly, I want to thank Dr. Stefan Slesazeck for continuous supervision, advice and countless insightful discussions on theoretical and practical problems related to negative capacitance. I also want to thank Dr. Uwe Schröder for many helpful discussions and support related to ferroelectric hafnium oxide.

My special thanks go out to Franz Fengler and Benjamin Max for fabricating some of the best samples I could have hoped for. Milan Pešić is acknowledged for the many interesting discussions (and beers) over the years. Furthermore, I want to thank all the other former and current members of the Ferro-Group at NaMLab which I had the pleasure to work with: Tony Schenk, Claudia Richter, Halid Mulaosmanovic, Terence Mittmann, Evelyn Breyer, Viktor Havel, Monica Materano, Min Hyuk Park, Furqan Mehmood and Patrick Lomenzo. Overall, I want to thank the whole NaMLab team, who make this institute what it is: One of the best places to work at.

I also want to thank Prof. Sayeef Salahuddin for creating this interesting field of research and for enabling my visit of his lab in 2016. These 6 months were incredibly inspiring and motivating for me. Furthermore, I want to thank Prof. Asif Khan for continued inspiration, collaboration and encouragement. I also want to thank my former colleagues at UC Berkeley: Korok Chatterjee, Claudy Serrao, Golnaz Karbasian, Samuel Smith and Zhongyuan Lu for creating an awesome work environment. I am also grateful to my collaborators at Globalfoundries Dresden, Raluca Negrea and Lucian Pintilie from NIMP, Uwe Mühle from Fraunhofer IKTS as well as Qinghua Han from Forschungszentrum Jülich.

Last but not least, I want to thank my friends and family for their emotional support during all these years. Especially, I want to thank Maxi and Melina for always being there for me. Lastly, I am deeply thankful for my parents, who have always supported me.

Curriculum vitae

Name:	Michael Hoffmann
Date of birth:	01.01.1990
Place of birth:	Neubrandenburg, Germany
Nationality:	German
E-mail address:	michaelhoffmann90@web.de

Education

2016/09 - now — **Ph.D. candidate in Electrical Engineering**
Faculty of Electrical and Computer Engineering, TU Dresden, Germany
Thesis: *"Negative Capacitance in Ferroelectric Materials"*

2013/10 - 2016/01 — **M.Sc. in Electrical Engineering**
Faculty of Electrical and Computer Engineering, TU Dresden, Germany
Thesis: *"Reliability Characterization and Modeling of State-of-the-Art High-k Metal Gate and Ferroelectric Field-Effect Transistors"*

2009/10 - 2013/09 — **B.Sc. in Electrical Engineering and Information Technology**
Otto-von-Guericke University Magdeburg, Germany
Thesis: *"Optimization of atomic layer deposition tool parameters and processes by analyzing Al_2O_3 depositions"*

Research experience

2016/09 - now — Scientist at NaMLab gGmbH, Dresden, Germany
Ferroelectricity and negative capacitance in HfO_2 and ZrO_2 based materials

2016/02 - 2016/07 — Visiting Student Researcher at University of California, Berkeley, USA
Characterization and modeling of ferroelectric negative capacitance

2013/12 - 2015/07 — Research Assistant at NaMLab gGmbH, Dresden, Germany
HfO_2 and ZrO_2 based ferroelectric thin films

2013/03 - 2013/09 — Research Assistant at Otto-von-Guericke University Magdeburg, Germany
Atomic layer deposition of Al_2O_3 and B_2O_3 thin films

2012/10 - 2013/02 — Internship at GLOBALFOUNDRIES Dresden, Germany
28 nm high-k metal gate stack reliability and electrical characterization

2011/12 - 2012/09 — Research Assistant at Otto-von-Guericke University Magdeburg, Germany
Atomic layer deposition of Al_2O_3 thin films

List of publications

A. Journal articles

1. **M. Hoffmann** and T. Mikolajick. Verborgene Energielandschaften. *Physik in unserer Zeit*, in press.

2. S. Pentapati, R. Perumal, S. Khandelwal, **M. Hoffmann**, S. K. Lim and A. I. Khan. Cross-domain optimization of ferroelectric parameters for negative capacitance transistors—Part I: Constant supply voltage. *IEEE Transactions on Electron Devices*, 67: 365–370, 2019.

3. **M. Hoffmann**, P. V. Ravindran and A. I. Khan. Why Do Ferroelectrics Exhibit Negative Capacitance? *Materials*, 12: 3743, 2019.

4. **M. Hoffmann**, F. P. G. Fengler, B. Max, U. Schroeder, S. Slesazeck, and T. Mikolajick. Negative Capacitance for Electrostatic Supercapacitors. *Advanced Energy Materials*, 9: 1901154, 2019.

5. **M. Hoffmann**, F. P. G. Fengler, M. Herzig, T. Mittmann, B. Max, U. Schroeder, R. Negrea, P. Lucian, S. Slesazeck, and T. Mikolajick. Unveiling the double-well energy landscape in a ferroelectric layer. *Nature*, 565: 464–467, 2019.

6. F. Mehmood, **M. Hoffmann**, P. D. Lomenzo, C. Richter, M. Materano, U. Schroeder and T. Mikolajick. Bulk Depolarization Fields as a Major Contributor to the Ferroelectric Reliability Performance in Lanthanum Doped $Hf_{0.5}Zr_{0.5}O_2$ Capacitors. *Advanced Materials Interfaces*, 1901180, 2019.

7. B. Max, **M. Hoffmann**, S. Slesazeck and T. Mikolajick. Direct Correlation of Ferroelectric Properties and Memory Characteristics in Ferroelectric Tunnel Junctions. *IEEE Journal of the Electron Devices Society*, 2019.

8. U. Schroeder, C. Richter, M. H. Park, T. Schenk, M. Pesic, **M. Hoffmann**, F. Fengler, D. Pohl, B. Rellinghaus, C. Zhou, C.-C. Chung, J. Jones and T. Mikolajick. Lanthanum doped hafnium oxide: a robust ferroelectric material. *Inorganic Chemistry*, 57: 2752–2765, 2018.

9. T. Schenk, **M. Hoffmann**, M. Pešić, M. H. Park, C. Richter, U. Schroeder, and T. Mikolajick. Physical Approach to Ferroelectric Impedance Spectroscopy: The Rayleigh Element. *Phys. Rev. Applied*, 10: 064004, 2018.

10. M. Pešić, T. Li, V. Di Lecce, **M. Hoffmann**, M. Materano, C. Richter, B. Max, S. Slesazeck, U. Schroeder, L. Larcher and T. Mikolajick. Built-in Bias Generation in Anti-ferroelectric Stacks: Methods and Device Applications. *IEEE Journal of the Electron Devices Society*, 6: 1019–1025, 2018.

11. M. H. Park, C.-C. Chung, T. Schenk, C. Richter, **M. Hoffmann**, S. Wirth, J. L. Jones, T. Mikolajick and U. Schroeder. Origin of temperature-dependent ferroelectricity in Si-doped HfO_2. *Adv. Electr. Mater.*, 4: 1700489, 2018.

12. **M. Hoffmann**, M. Pešić, S. Slesazeck, U. Schroeder and T. Mikolajick. On the Stabilization of Ferroelectric Negative Capacitance in Nanoscale Devices. *Nanoscale*, 10: 10891–10899, 2018.

13. **M. Hoffmann**, A. I. Khan, C. Serrao, Z. Lu, S. Salahuddin, M. Pešić, S. Slesazeck, U. Schroeder and T. Mikolajick. Ferroelectric Negative Capacitance Domain Dynamics. *J. Appl. Phys.*, 123: 184101, 2018.

14. Q. Han, T. C. U. Tromm, **M. Hoffmann**, P. Aleksa, U. Schroeder, J. Schubert, S. Mantl and Q.-T. Zhao. Subthreshold Behavior of Floating Gate MOSFETs with Ferroelectric Capacitors. *IEEE Transactions on Electron Devices*, 65: 4641–4645, 2018.

15. F. P. G. Fengler, **M. Hoffmann**, S. Slesazeck, T. Mikolajick and U. Schroeder. On the relationship between field cycling and imprint in ferroelectric $Hf_{0.5}Zr_{0.5}O_2$. *J. Appl. Phys.*, 123: 204101, 2018.

16. M. Pešić, C. Künneth, **M. Hoffmann**, H. Mulaosmanovic, S. Müller, E. T. Breyer, U. Schroeder, A. Kersch, T. Mikolajick and S. Slesazeck. A computational study of hafnia-based ferroelectric memories: from ab initio via physical modeling to circuit models of ferroelectric device. *Journal of Computational Electronics*, 16: 1236–1256, 2017.

17. M. H. Park, T. Schenk, **M. Hoffmann**, S. Knebel, J. Gärtner, T. Mikolajick and U. Schroeder. Effect of Acceptor Doping on Phase Transitions of HfO_2 Thin Films for Energy-Related Applications. *Nano Energy*, 36: 381-389, 2017.

18. A. I. Khan, **M. Hoffmann**, K. Chatterjee, Z. Lu, R. Xu, C. Serrao, S. Smith, L. W. Martin, C. C. Hu, R. Ramesh, S. Salahuddin. Differential voltage amplification from ferroelectric negative capacitance. *Appl. Phys. Lett.*, 111: 253501, 2017.

19. **M. Hoffmann**, T. Schenk, M. Pešić, U. Schroeder and T. Mikolajick. Insights into antiferroelectrics from first-order reversal curves. *Appl. Phys. Lett.*, 111: 182902, 2017.

20. M. Pešić, **M. Hoffmann**, C. Richter, T. Mikolajick and U. Schroeder. Nonvolatile Random Access Memory and Energy Storage Based on Antiferroelectric Like Hysteresis in ZrO_2. *Adv. Funct. Mater.*, 26: 7486–7494, 2016.

21. **M. Hoffmann**, M. Pešić, K. Chatterjee, A. I. Khan, S. Salahuddin, S. Slesazeck, U. Schroeder and T. Mikolajick. Direct Observation of Negative Capacitance in Polycrystalline Ferroelectric HfO_2. *Adv. Funct. Mater.*, 26: 8643–8649, 2016.

22. T. Schenk, **M. Hoffmann**, J. Ocker, M. Pešić, T. Mikolajick and U. Schroeder. Complex internal bias fields in ferroelectric hafnium oxide. *ACS Appl. Mater. Interfaces*, 7: 20224–20233, 2015.

23. **M. Hoffmann**, U. Schroeder, C. Künneth, A. Kersch, S. Starschich, U. Böttger and T. Mikolajick. Ferroelectric phase transitions in nanoscale HfO_2 films enable giant pyroelectric energy conversion and highly efficient supercapacitors. *Nano Energy*, 18: 154–164, 2015.

24. **M. Hoffmann**, T. Schenk, I. Kulemanov, C. Adelmann, M. Popovici, U. Schroeder and T. Mikolajick. Low Temperature Compatible Hafnium Oxide Based Ferroelectrics. *Ferroelectrics*, 480: 16–23, 2015.

25. **M. Hoffmann**, U. Schroeder, T. Schenk, T. Shimizu, H. Funakubo, O. Sakata, D. Pohl, M. Drescher, C. Adelmann, R. Materlik, A. Kersch and T. Mikolajick. Stabilizing the ferroelectric phase in doped hafnium oxide. *J. Appl. Phys.*, 118, 072006, 2015.

B. Conference contributions

1. **M. Hoffmann**, P. V. Ravindran and A. I. Khan. (Invited Talk) A Microscopic "Toy" Model of Ferroelectric Negative Capacitance. *Electron Devices Technology and Manufacturing (EDTM) Conference*, Penang, Malaysia, March 2020.

2. T. Mikolajick, U. Schroeder, P. D. Lomenzo, E. T. Breyer, H. Mulaosmanovic, **M. Hoffmann**, T. Mittmann, F. Mehmood, B. Max and S. Slesazeck. (Invited) Next Generation Ferroelectric Memories enabled by Hafnium Oxide. *International Electron Device Meeting (IEDM)*, San Francisco, CA, USA, December 2019.

3. P. D. Lomenzo, S. Slesazeck, **M. Hoffmann**, B. Max, T. Mikolajick and U. Schroeder. (Invited) Ferroelectric $Hf_{1-x}Zr_xO_2$ Memories: Device Reliability and Depolarization Fields. *Non-Volatile Memory Technology Symposium (NVMTS)*, Durham, NC, USA, October 2019.

4. T. Mikolajick, T. Mittmann, **M. Hoffmann**, B. Max, H. Mulaosmanovic, M. H. Park, S. Slesazeck and U. Schroeder. (Invited) Basics and Device Applications of Ferroelectricity in Hafnium Oxide. *International Symposium on Integrated Functionalities (ISIF)*, Dublin, Ireland, August 2019.

5. S. Slesazeck, H. Mulaosmanovic, **M. Hoffmann** and T. Mikolajick. (Invited) Switching kinetics in Hafnium oxide based ferroelectric/dielectric bilayer stacks. *Insulating films on semiconductors (INFOS) conference*, Cambridge, UK, July 2019.

6. **M. Hoffmann**, S. Slesazeck, and T. Mikolajick. Dynamic modeling of hysteresis-free negative capacitance in ferroelectric/dielectric stacks under fast pulsed voltage operation. *Device Research Conference (DRC)*, Ann Arbor, MI, USA, June 2019.

7. F. Winkler, M. Pešić, C. Richter, **M. Hoffmann**, T. Mikolajick and J. W. Bartha. Demonstration and Endurance Improvement of p-channel Hafnia-based Ferroelectric Field Effect Transistors. *Device Research Conference (DRC)*, Ann Arbor, MI, USA, June 2019.

8. T. Mikolajick, H. Mulaosmanovic, **M. Hoffmann**, B. Max, T. Mittmann, U. Schroeder and S. Slesazeck. Variants of Ferroelectric Hafnium Oxide based Nonvolatile Memories. *Device Research Conference (DRC)*, Ann Arbor, MI, USA, June 2019.

9. **M. Hoffmann**. (Invited Talk) Negative capacitance effects in HfO_2 based ferroelectrics. *Novel high-k Application Workshop*, Dresden, Germany, June 2019.

10. B. Max, **M. Hoffmann**, S. Slesazeck and T. Mikolajick. Retention Characteristics of $Hf_{0.5}Zr_{0.5}O_2$-based Ferroelectric Tunnel Junctions. *International Memory Workshop (IMW)*, Monterey, CA, USA, May 2019.

11. S. Slesazeck, V. Havel, E. T. Breyer, H. Mulaosmanovic, **M. Hoffmann**, B. Max, S. Duenkel and T. Mikolajick. Uniting The Trinity of Ferroelectric HfO2 Memory Devices in a Single Memory Cell. *International Memory Workshop (IMW)*, Monterey, CA, USA, May 2019.

12. U. Schroeder, T. Mikolajick, **M. Hoffmann**, B. Max and S. Slesazeck. (Invited) Negative Capacitance in Ferroelectric Hafnium Oxide. *MRS Spring Meeting*, Phoenix, AZ, USA, April 2019.

13. **M. Hoffmann**, B. Max, T. Mittmann, U. Schroeder, S. Slesazeck and T. Mikolajick. Demonstration of High-speed Hysteresis-free Negative Capacitance in Ferroelectric $Hf_{0.5}Zr_{0.5}O_2$. *International Electron Device Meeting (IEDM)*, San Francisco, CA, USA, December 2018.

14. B. Max, **M. Hoffmann**, S. Slesazeck and T. Mikolajick. Ferroelectric Tunnel Junctions based on Ferroelectric-Dielectric $Hf_{0.5}Zr_{0.5}O_2 / Al_2O_3$ Capacitor Stacks. *European Solid-State Device Research Conference (ESSDERC)*, Dresden, Germany, September 2018.

15. **M. Hoffmann**, S. Slesazeck, and T. Mikolajick. Domain Formation in Ferroelectric Negative Capacitance Devices. *Device Research Conference (DRC)*, Santa Barbara, CA, USA, June 2018.

16. A. Aziz, E. T. Breyer, A. Chen, X. Chen, S. Datta, S. K. Gupta, **M. Hoffmann**, X. S. Hu, A. Ionescu, M. Jerry, T. Mikolajick, H. Mulaosmanovic, K. Ni, M. Niemier, I. O' Connor, A. Saha, S. Slesazeck, S. K. Thirumala, X. Yin. Computing with Ferroelectric FETs: Devices, Models, Systems, and Applications. *Design, Automation and Test in Europe (DATE) conference*, Dresden, Germany, March 2018.

17. **M. Hoffmann** and S. Salahuddin. (Invited Talk) Negative Capacitance Transistors. *Design, Automation and Test in Europe (DATE) conference*, Dresden, Germany, March 2018.

18. M. Pešić, **M. Hoffmann**, S. Slesazeck, T. Mikolajick, V. Di Lecce and L. Larcher. HfO_2 based Ferroelectric Memories: Physical and Circuit Modeling of Ferroelectric Devices. *IEEE SOI-3D-Subthreshold Microelectronics Technology Unified Conference*, San Francisco, CA, USA, October 2017.

19. M. H. Park, T. Schenk, **M. Hoffmann**, S. Knebel, J. Gaertner, T. Mikolajick and U. Schroeder. Phase Transitions in Doped Hafnia Thin Films for Pyroelectric Applications. *The 14th International Meeting on Ferroelectricity*, San Antonio, TX, USA, September 2017.

20. M. Pešić, **M. Hoffmann**, C. Richter, S. Slesazeck, T. Kämpfe, L. M. Eng, T. Mikolajick and U. Schroeder. Anti-ferroelectric ZrO_2: An Enabler for Low Power Non-volatile 1T-1C and 1T Random Access Memories. *European Solid-State Device Research Conference (ESSDERC)*, Leuven, Belgium, September 2017.

21. M. Pešić, **M. Hoffmann**, C. Richter, S. Slesazeck, U. Schroeder and T. Mikolajick. Anti-ferroelectric-like ZrO_2 non-volatile memory: Inducing non-volatility within state-of-the-art DRAM. *Non-Volatile Memory Technology Symposium (NVMTS)*, Aachen, Germany, August 2017.

22. U. Schroeder, T. Schenk, **M. Hoffmann**, C. Richter, M. Pesic, F. Fengler, T. Mittmann, S. Slesazeck, A. Kersch, J. L. Jones, J. LeBeau, J. Müller, S. Müller and T. Mikolajick. Applications

for Ferroelectric Hafnium Oxide: From non-volatile memory to steep slope devices. *Insulating films on semiconductors (INFOS) conference*, Potsdam, Germany, June 2017.

23. M. Pešić, **M. Hoffmann**, C. Richter, S. Slesazeck, T. Mikolajick and U. Schroeder. Anti-ferroelectric HfO_2 or ZrO_2: A key material for novel anti-ferroelectric non-volatile memories. *IEEE International Symposium on Applications of Ferroelectric (ISAF)*, Atlanta, GA, USA, May 2017.

24. U. Schroeder, T. Schenk, **M. Hoffmann**, C. Richter, M. Pesic, F. Fengler, S. Slesazeck, S. Kalinin, A. Kersch, J. L. Jones, J. LeBeau and T. Mikolajick. (Invited) Ferroelectric HfO_2 or ZrO_2 for Non-Volatile Memory Devices. *MRS Spring Meeting*, Phoenix, AZ, USA, April 2017.

25. T. Schenk, M. H. Park, M. Pešić, **M. Hoffmann**, C. Richter, S. Mueller, H. Mulaosmanovic, F. P. G. Fengler, S. Slesazeck, T. Mikolajick and U. Schroeder. 10 Years Fluorite-type Ferroelectrics – A Survey. *Towards oxide-based Electronics Spring Meeting*, Luxembourg, April 2017.

26. **M. Hoffmann**, M. Pešić, S. Slesazeck, U. Schroeder and T. Mikolajick. Modeling and Design Considerations for Negative Capacitance Field-Effect Transistors. Joint International EUROSOI Workshop and International Conference on Ultimate Integration on Silicon (ULIS), Athens, Greece, April 2017.

27. **M. Hoffmann**. (Invited Talk) Modeling and Design Considerations for Negative Capacitance Field-Effect Transistors. *Novel high-k Application Workshop*, Dresden, Germany, March 2017.

28. M. Pešić, S. Knebel, **M. Hoffmann**, C. Richter, T. Mikolajick and U. Schroeder. How to Make DRAM non-volatile? Anti-ferroelectrics: A New Paradigm for Universal Memories. *International Electron Device Meeting (IEDM)*, San Francisco, CA, USA, December 2016.

29. T. Schenk, **M. Hoffmann**, J. Ocker, M. Pešić, E. D. Grimley, X. Sang, J. M. LeBeau, T. Mikolajick and U. Schroeder. Internal Bias Fields in Ferroelectric HfO_2 Thin Films and their Structural Origins. *IEEE International Symposium on Applications of Ferroelectric (ISAF)*, Darmstadt, Germany, August 2016.

30. U. Schroeder, T. Schenk, **M. Hoffmann**, C. Richter, M. Pešić, F. Fengler, M.H. Park, S. Slesazeck, D. Pohl, C. Künneth, R. Materlik, A. Kersch, X. Sang, E. Grimley, J. LeBeau, N. Wisinger, S. Kalinin and T. Mikolajick. (Invited) Ferroelectric HfO_2 for Non-Volatile Memory Devices. *International Conference Smart and Multifunctional Materials, Structures and Systems (CIMTEC)*, Perugia, Italy, June 2016.

31. T. Schenk, **M. Hoffmann**, C. Richter, M. Pešić, S. Mueller, S. Slesazeck, U. Schroeder, T. Mikolajick, D. Pohl, J. Mueller, P. Polakowski, R. Materlik, A. Kersch, X. Sang, E. D. Grimley and J. M. LeBeau. (Invited) Doped Hafnium Oxide for Ferroelectric Memories. *MRS Fall Meeting*, Boston, MA, USA December 2015.

32. U. Schroeder, T. Schenk, C. Richter, **M. Hoffmann**, D. Martin, R. Materlik, A. Kersch, D. Pohl, S. Kalinin and T. Mikolajick. (Invited) Piezo-, Pyro, and Ferroelectric Properties of ALD HfO_2 Based Nanolaminates. *Baltic ALD conference*, Tartu, Estonia, September 2015.

33. U. Schroeder, T. Schenk, **M. Hoffmann**, C. Richter, R. Materlik, A. Kersch, M. H. Park, C. S.

Hwang, D. Pohl, S. Kalinin and T. Mikolajick. Ferroelectric HfO_2 for Novel Semiconductor Devices. *International Conference on Atomic Layer Deposition*, Portland, Oregon, USA, July 2015.

34. U. Schroeder, T. Schenk, C. Richter, **M. Hoffmann**, D. Martin, T. Shimizu, H. Funakubo, D. Pohl, C. Adelmann, R. Materlik, A. Kersch, X. Sang, J. LeBeau, N. Wisinger, A. Ievlev, S. Kalinin and T. Mikolajick. (Invited) Searching for the Origin of the Ferroelectric Phase in HfO_2. *IEEE International Symposium on Applications of Ferroelectrics (ISAF)*, Singapore, May 2015.

35. T. Schenk, U. Schroeder, **M. Hoffmann**, M. Pešić, M. Popovici, Y. V. Pershin and T. Mikolajick. Insights into Analyzing the Electric Field Cycling Behavior of Ferroelectric Doped Hafnium Oxide. *European Conference on Application of Polar Dielectrics (ECAPD)*, Vilnius, Lithuania, July 2014.

C. Book chapters

1. S. Slesazeck, H. Mulaosmanovic, **M. Hoffmann**, U. Schroeder, B. Max and T. Mikolajick. Chapter 12: MO_x in Ferroelectric memories. Editors: Ilia Valov and Panagiotis Dimitrakis, *Metal oxides for Non-Volatile Memories: Materials, Technology and Applications*, in press.

2. **M. Hoffmann**, S. Slesazeck, T. Mikolajick and C. S. Hwang. Chapter 10.5 - Negative Capacitance in HfO_2- and ZrO_2-Based Ferroelectrics, Editors: Uwe Schroeder, Cheol Seong Hwang, Hiroshi Funakubo, *Ferroelectricity in Doped Hafnium Oxide: Materials, Properties and Devices*, Woodhead Publishing, 2019, Pages 473-493.

3. M. H. Park, **M. Hoffmann** and C. S. Hwang. Chapter 5.2 - Pyroelectric and Electrocaloric Effects and Their Applications, Editors: Uwe Schroeder, Cheol Seong Hwang, Hiroshi Funakubo, *Ferroelectricity in Doped Hafnium Oxide: Materials, Properties and Devices*, Woodhead Publishing, 2019, Pages 217-244.